Classic Cones

Pictorial Reference and Value Guide for 1920's Radio Cone Speakers

Buford and Jane Chidester

SONORAN PUBLISHING, LLC
Chandler, Arizona

Sonoran Publishing, LLC, Chandler, Arizona 85226

First printing December 2001

Printed in the United States of America

Library of Congress Cataloging-in-Publication Data

Chidester, Buford, 1942-

Classic cones : pictorial reference and value guide for 1920's radio cone speakers / Buford and Jane Chidester.

p. cm.

Includes index.

ISBN 1-886606-16-1 (pbk.)

1. Loud speakers--Collectors and collecting--Catalogs. 2. Radio--Equipment and supplies--Collectors and collecting--Catalogs. I. Chidester, Jane, 1940- II. Title.

TK5983 .C49 2001

621.384'13--dc21 2001054924

Dedication

First of all, I would like to dedicate this book to my parents, Orval and Icie Chidester. Nestled in the little village of Canaan, West Virginia, they were both very patient and supportive as I experimented my way through the early years of this hobby. Experimenting with such things as installing speaker wires from one end of the house to the other; connecting speaker wires directly into 120 VAC receptacles causing a power outage; and filling my father's spare truck battery with soda water, causing a geyser similar to Old Faithful. It's a miracle that I am alive today. Thinking back on those years, I can now see how beneficial it was for me to indulge in a learning process such as I went through at that time. My dad was a radio buff himself and was very generous in helping me to learn and understand the theory of radio. One of my first complete radios, a present from him that I enjoyed very much, was a Grebe MU1 with a Model 30 Utah ornate wooden speaker. Although I didn't know how to make the radio work, I spent many hours tinkering with it. Unfortunately, as fate would have it, curiosity got the best of me and I removed most of the parts trying to find what made the tubes shake and how those nice green coils were wound. It was always gratifying to see how the O1A's would glow when applying 6 or 12 VAC to their filaments. Of course, the tubes didn't last long either. This was the demise of a beautiful Grebe radio. The Utah speaker did survive to give me hours of pleasure while listening to the one tube radios that Dad help me put together. I have never forgotten those days and have found the knowledge, gained through those experiments, very useful throughout my entire electrical design career.

One other person that I must dedicate this book to, is my wife Jane (co-author). Without her help and encouragement that she has given me, I probably would not be involved in this book today. She has spent many hours researching endless leads and has literally let the genie out of the bottle, providing valuable information for this book.

Acknowledgements

Many collectors have freely given us help in procuring information regarding cone speakers. We certainly appreciate the help that we were given, allowing us to provide a reference book that will be a benefit to all collectors. Special mention of persons that have contributed pictures, ads and items, are acknowledged below. For those that we have failed to mention, we appreciate the support you have provided and hope you will benefit from the information presented here.

Leonard Aquino, Larry and Dot Babcock, Merrill and Rita Bancroft, R. L. and Linda Barnett, Ed Bell, Geoff Bourne and Grace MacMillen, Dick Bury, Gary and Laura Carter, Fred and Margurite Crews, Dave Crocker, John and Judy DeLoria, Alan Douglas, Terry Dowling, Richard Gorman, Greg and Mary Farmer, Dave and Julie Fischer, Herb Fredrickson, Glenn and Cookie Heimroth, Bill Holly, Marc Ingenthron, Wayne King, Arthur Kingsley, Sam and Susan Luscomb, Bruce and Charlotte Mager, Ted Miller, Michael Murray, Bill Overbeck, Roger Sannipoli, Ludwell Sibley, Bob Slagle, Charles Stinger, Steve Wallace, Jay Volke.

Introduction

Welcome to our world: the world of Magnetic Cone Speakers. Cone speakers were one of the many fascinating products of the 1920's era. This type speaker played a very important role in the development of radio and its accessories in the 1920's. They have been called the "Soul of Radio," and many, in their own right, are works of art. They can be romantic, patriotic, inspirational, and even sexy. Cone speakers could be purchased as table models; pedestal, lamp stand or console floor models, and wall mounted units. Designs included ships, airplanes, castles, butterflies, cathedrals, dancing girls and many other picturesque scenes and geometric designs that were either cast into the speaker frame or portrayed on the cone. Tower, Pathe', O'Neil, and others, supplied cone speakers with optional lights that provided background lighting. Speakers were also combined with table lamps or chandeliers. The Miniature Ship Company produced a series of solid wooden ship speakers, ranging in size from 25 to 50 inches in length. Several models included lights that were mounted in the hull for illuminating multicolor portholes and other translucent objects on the ship's deck.

We would like to mention some events that were important in leading up to the development of cone speakers. The 1920's enjoyed very high inventive activity in the development of radio that was to evolve into the vast arena of communications we have today. In the early 1920's, radio was in its infancy as far as the development of radios that were suitable for home use. Low power sets, such as the crystal and one or two tube radio sets, were the only type available. They used a variety of earphones for sound output. Generally, earphones allowed only one person to listen to the radio at a time. Sound output of this type of device was very small. In some instances, earphones were placed in a bowl or similar object, reflecting sound, so more that just one person could listen.

Another alternative to the earphone, the horn speaker, had been invented a few years earlier. Horn speakers required higher power for operation and were not widely used until more powerful radio sets became available. The radio industry saw an increasing demand for loudspeakers starting in the early 1920's. As the industry grew, more powerful transmitters and receivers were developed, allowing use of higher volume loudspeakers. Although horn speakers had more volume, they also had inherent drawbacks. One of the obvious problems with most horn speakers was their inability to reproduce a wide range of high and low notes. Horn speaker designs allowed only a narrow band of audio frequencies, due mostly to the small diaphragms used for producing the sound. The majority of horns were manufactured from hard rubber, concrete, wood, papier-mache, and various other sturdy materials. Although some horn speakers were manufactured from beautiful wood or colorful materials, such a pyralin, most didn't really add to the décor of their surroundings.

As radio popularity grew and designs were improved, a new type speaker, advertised as the cone reproducer, was introduced to the market in early 1924. Cone speakers solved many of the problems that were inherent in earlier earphone and horn speaker designs. One of the major advantages of the cone speaker was better sound reproducing capability. Unlike the horn speakers with very small diaphragms, cone speakers were manufactured with cone sizes ranging from approximately 4 to 36 inches in diameter. The larger diameter cones allowed a broad range of audible frequencies. A wide selection of cone speakers were available allowing customers to purchase a speaker of their choice. Cone speakers were produced in various designs, colors, shapes and sizes. Like the high tech industry of today, large numbers of manufacturers, job shops, and distributors were bringing to market new designs, ideas, and gimmicks to get their share of business. Because of this fierce competition, vast improvements were made to the design of the cone speaker, allowing it to become the front runner in loudspeaker sales.

Cone speakers had their heyday in the short period from 1924 into 1929 and also became a victim of another new era when electric radios were produced with speakers integrated into radio cabinets. In the author's opinion, this short period had the largest variety and most beautiful of all radio speakers that have ever been produced for the general public.

Organization

Where possible, the speaker manufacturer is shown in alphabetical order. In some instances, speakers were name-branded and distributed by many different outlets without actually representing the original manufacturer. Speakers are also listed in alphabetical order under a distributor or speaker name where the actual manufacturer was not known. Speakers that have not been identified or are of other types such as horn, air column or condenser, may appear in the back of this book under the title "Other Speakers of Interest." It is almost impossible to identify some speakers by comparing physical characteristics. We have tried to document all speakers in this book as accurately as possible. There are many variations in color and design that also exist but are not pictured. It is also important to note that identical speakers can be found with various distributor or manufacturer labels.

Advertisements contained in this book are from newspapers, books, magazines, and catalogs dated from 1923 through 1930 and may not represent the exact year of manufacture. Where possible, we have given credit to the publication where we have obtained the documentation. Patent dates on speaker nameplates should not be used to determine the age of the speaker. Many speaker concepts were patented years before actual manufacturing began. Some ads may show speakers that may not have been manufactured. Some ads may show speakers that were never manufactured and thought to be for marketing purposes only. Some speakers may not have been produced for reasons such as companies changing their marketing strategies or actually going out of business.

Speaker sizes are listed for reference only. The height is normally shown, but in some instances the height and width are shown where we felt necessary. Some speakers are marked "Approx." for those that were not available for actual measurement.

Speaker Values

Values in this book have been derived from information obtained at radio meets, flea markets, antique shops, auctions, and personal experience. Values are listed for speakers that have good cosmetic appearance and are in working order. Speakers of this type usually must have some restoration such as cord replacement, cone, or driver repairs. Any repairs or restoration must be as close to the original design, color, and materials as feasible. Restoration should not degrade the speaker's original appearance or performance. We believe restoration in this context is acceptable within reason. Values can vary widely for many factors such as condition, manufacturer, desirability, regional demand, and availability. Speakers that are considered new and are in original shipping cartons, or are in unusually fine condition may be priced well above the value shown. In some instances, values have not been listed due to their rarity, and will be labeled "Rare". However, rare does not necessarily mean higher value: it could mean the speaker is just not available. Speakers marked "Rare" usually are in private collections and are not sold within the general market, preventing reasonable pricing to be determined. Values for rare speakers are generally agreed upon by the immediate buyer and seller and may be well above the normal market values of the more common speakers.

Acme Junior Model K3
Table model with metal base and frame. Green and gold crackle finish. Brown and green marbled cone.
H=14″ $150-175

Radio News, Dec. 1926

ACME FREE EDGE CONE SPEAKERS

K-3

A "single free-edge cone" of eleven inches diameter. Faithfulness of reproduction is maintained in this unit. Actuating unit is of rugged construction. Plate currents up to 25 MA can be safely used thereby providing great volume, if desired.

The K-3 speaker can be adapted to wall mounting by simply removing the base and adding a special fitting to which a silk cord can be attached. An adjustment is provided on the back for increased plate current when used. The metal case and base are finished in green bronze of beautiful color.

102PH5149— Price, prepaid........ **$16.65**

K-2

Acme Cone Speaker in Cabinet

Same as type K-1, except encased in a piano finished solid mahogany cabinet. Fits in well with the furnishings of any home and matches very well with the cabinets of most receiving sets. May b[illegible] set upon the cabinet of table or floor consoles without the usual microphonic noises.

102PH5148—Acme Type K-2. Price, prepaid **$34.20**

K-1

The Acme K-1 "double free-edge cone speaker" appeared on the market last year for the first time and has met with tremendous success. With the Acme K-1 on a set, notes, tones, and inflection are brought out as never before and for the first time, orchestra volume can be had without the desire to turn it down. The units used in Acme speakers are entirely new and of such rugged construction that permanency is assured. The cones are each 14 inches in diameter, giving a large radiating surface. The use of the free-edge cone insures a flat frequency range and prevents any detrimental warping or shrinking due to moisture or changes in weather.

The direct current in the speaker circuit aids in the operation and nee[illegible] not be shunted out as in some speakers. Values up to 25 MA [illegible] safely be used, allowing the speaker to be directly conn[illegible] to power tubes. A single adjustment is provided for [illegible] care of increased plate current when used. The m[illegible] is finished in tan and easily harmonizes with all kinds [illegible]ng room furniture.

102PH5147—Price, prepaid................ **$22.50**

Acme Model K1
Table model with double free edge cone. Metal frame with tan speckled finish. Optional marbled cone. Cone front and back.
H=14" $80-90

Acme Model K1
Table model with double free edge cone. Metal frame with mahogany finish. Optional marbled cone. Cone front and back.
H=14" $80-90

NEW AIR-CHROME SPEAKERS

AIR-CHROME MODEL F

Beautiful Solid Mahogany Frame. Width 18 in., height 27¾ in., depth 9 in. A floor model at a reasonable price. Ideal for the small living room.

List Price $35.00

Net Price, $21.00

AIR-CHROME MODEL G

American Walnut Console. Width 28¾ in., height 31 in., depth 14½ in. One of the most beautiful of the smaller consoles.

List Price $65.00

Net Price, $39.00

AIR-CHROME MODEL K

This is Air-Chrome's latest addition to their already complete line of speakers. Combines the superior Air-Chrome diaphragm with the dynamic unit. Encased in a beautiful Walnut Cabinet with carved decorations and fret-work. A.C. Volt Type.

List Price $75.00

Net Price, $45.00

AIR-CHROME MODEL J

American Walnut Cabinet — beautifully carved. The attractiveness of this model lies in its extreme simplicity of design and well proportioned size.

List Price $35.00

Net Price, $21.00

BOX MODELS FOR SET BUILDERS

24 in. x 24 in. List..... $25.00
Net Price, $15.00

14 in. x 14 in. List..... $20.00
Net Price, $12.00

18 in. x 23 in. List..... $23.00
Net Price, $13.80

NEW TEMPLE SPEAKERS

TEMPLE DYNAMIC No. 50

Model 50 is beautiful in appearance. Its rich and pleasing walnut panels give it an eye value that is acceptable to all. Operates on either A.C. or D.C., or battery.

Temple Model 50 Dynamic is 14¾ in. high, 11 in. wide, and 6¾ in. deep.

List Price $49.00

Net Price, $29.40

TEMPLE AIR-CHROME No. 20

Temple Air-Chrome is new in principle, marvelous in tone and beautiful in appearance; balanced tension diaphragm, volume without distortion. Its open radiator consists of two sections, instead of one, driven by the powerful Temple Double Action Unit, thus substantially increasing the volume ond tonal range. Temple Model 20 Air-Chrome is 14¾ in. high, 11 in. wide and 5¼ in. deep. Housed in same cabinet as Dynamic mentioned above.

List Price $29.00

Net Price, $17.40

TEMPLE AIR COLUMN No. 15
60 inch Centre Line

From the standpoint of appearance Model 15 will impress everyone—truly a masterpiece of harmony in its artitic and dignified lines. Even the most critical will acknowledge its faultless design. It is encased in genuine walnut with sides of the same color in beautiful grained leather effect.

Temple Air Column Model 15 is 11⅝ in wide, 12¼ in. high, and 8⅛ in. deep.

List Price $29.00

Net Price, $17.40

TEMPLE AIR-CHROME BOX SPEAKERS

MODEL Z

Model Z. The small, rectangular model. Wonderful tone—small in size. Size, 9¾ in. x 21 in., depth 5 in. List Price $20.00

Net Price, $12.00

MODEL T

Model T. This model is almost square. Size 10 in. x 13 in. The performance is the same as Model Z, the difference being in size only to allow installation in different types of cabinets. Depth 4 inches. List Price $20.00

Net Price, $12.00

TEMPLE COMPARATOR

The Temple Comparator allows the comparing of from two to five speakers. A slight turn of the knob switches from one speaker to another. May also be used as a selector for various speakers when they are located in different rooms.

List Price .... Each, $3.50

Net Price, $2.10

TEMPLE DOUBLE ACTION UNITS

List Price...............$10.00 Net Price...............$6.00

ROYAL - EASTERN ELECTRICAL SUPPLY COMPANY, 16 WEST 22ND STREET, NEW YORK

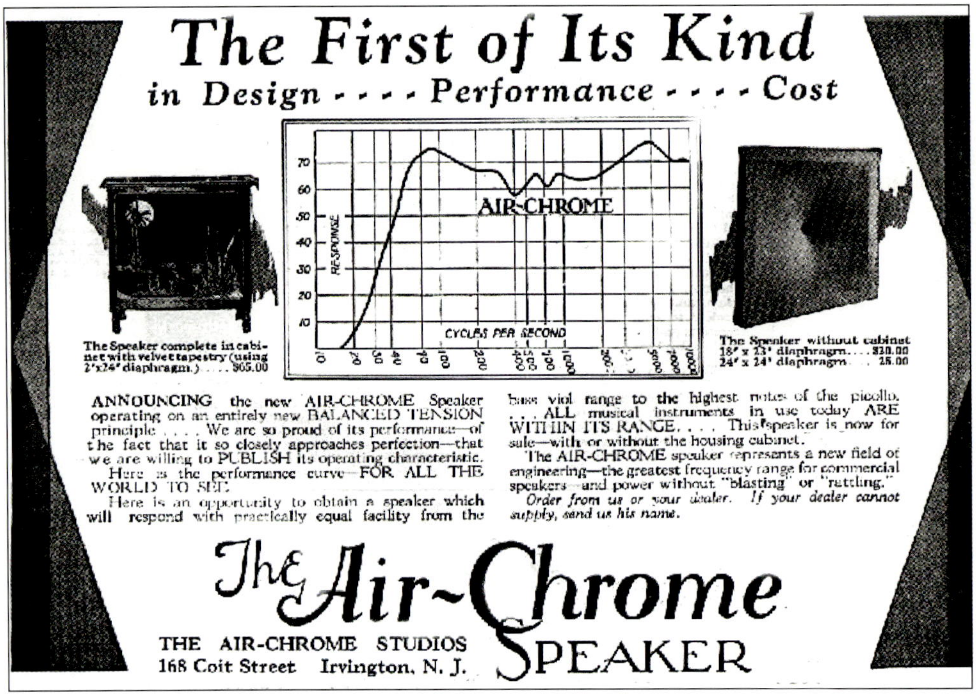

Radio News, March 1928

Air-Chrome

Console with oriental design silk front and sides. Double cones.

Approx. 30" square, Depth = 10" "Rare"

Air-Chrome

Floor table model with double cone mounted vertically inside.

H=30" 18" Square "Rare"

Air-Chrome
Table model with wood frame and double fabric cones.
H=14" $450-500

Air-Chrome
Table or wall hanging model with wood frame and double fabric cones.
H=Approx. 18" $150-175

Air-Chrome
Table or wall hanging model with bird and flower design.
H=Approx. 20" $500-600

Air-Chrome

Octagon table models . These speakers were decorated with many different designs and colors. Both have plaster ornate trim with wood base and open back. The forest scene is painted directly on the cone material and is signed by the artist. Whether this design was painted at the factory or at a later date is unknown. Many speakers were painted at the factory but many were also painted after leaving the factory. Speakers with grille cloth, such as the dragon shown, usually were either silk or tapestry. Depending on the cone design, prices could vary widely.

H=Approx. 20" $150-175

Temple Air-Chrome Model 20

Table model with walnut case and ornate front grille.

H=25", Width 20", Depth 10" $300-350

Barwicks Catalog, 1929

A New Triumphant Achievement In Speaker Performance

DISCOUNT 40 PER CENT
GRAYMORE RADIO CORP.
142 LIBERTY ST., NEW YORK CITY

AKRADYNE engineers have set a new standard in speaker construction . . . a new triumphant achievement in speaker performance . . . in their creation of Model No. 30.

The frequency range of Model 30 is from 50 to 5000 cycles, covering the entire frequency spectrum. Windings are constructed to provide the maximum amount of inductance with minimum distributed capacity, assuring a high degree of response on the higher audio frequencies. The inductance value of these windings is such that the speaker also responds exceptionally well on 50 cycles, bringing in the low notes so necessary to pleasing musical reproduction.

To protect "fans" who do not use speaker coupling devices, inductance windings have been designed to carry a large amount of current. Another protection is achieved in the speaker magnets, in which the highest quality steel is used to assure magnetic longevity.

The design of the speaker unit has been greatly improved, and with larger spacing between pole pieces and stylus, the speaker handles a large amount of power entirely without distortion or loss of sensitivity.

Model No. 30 is equipped with a unit eminently suited to handle the present day power tubes and will work satisfactorily on sets delivering either A.C. or D.C. output. It is truly a universal speaker.

Of sturdy, simple construction; richly colored to blend with furnishings found in modern living rooms . . . Model 30 is pleasing to eye as well as ear. Cone is approximately 24 in. in diameter, with a height from base of tripod of about 25 in. Table mounting is provided by this handsome claw-foot base.

Akradyne SPEAKER

MODEL No. 30

$30

GRAYMORE RADIO CORPORATION, 142 LIBERTY ST., N. Y. C.

Akradyne
Table model with wood frame and fabric single free edge cone.
H=Approx. 14.5″ **$90-110**

Speaker nameplate

Front

Back

Akradyne Model 30
Table model with cast iron stand and claw feet. Double free edge 22" cone. Re-inforced edge.
H=25″ **$250-300**

Algonquin

Table model with cast iron gold metal stand and frame stippled in black. Laminated gold fabric cone. Cone usually faded to olive green.
H=18.5″ $350-375

Algonquin

Table model with metal base and wood sounding board. Finish is gold stippled in black with a center cast iron ship. Laminated gold fabric cone. Cone usually faded to an olive green.
H=18.5″ $350-400

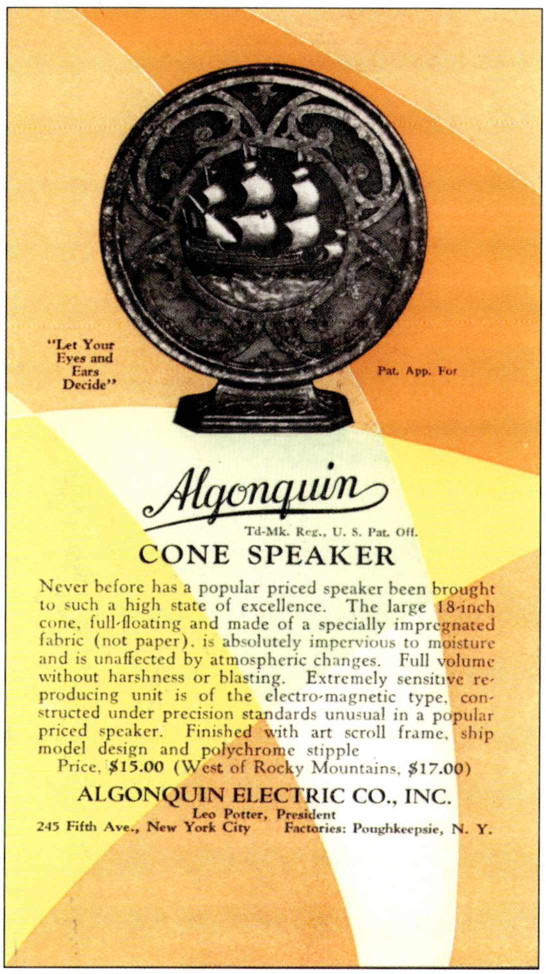

Algonquin speaker brochure

Algonquin speaker brochure

Radio, Sept. 1926

Haskell Radio Corp. Catalog

Front

Back

American Bosch Model 608

Table model cast iron speaker found in a variety of cone designs as shown. Original designs consist of a black /grey background with bronze and gold pictured at left, black/grey wrinkle finish with a relief of gold, shown at right, and the beautiful black/grey with dragon design. Base and stand are cast iron. Cone supported with a pressed metal ring. Base and stand finished in a rough wrinkle gold/bronze usually faded to a darker color.
H=19.5" $400-500

American Bosch Model 608

Table model with dragon design. Single free edge cone with metal tower stand. Base and stand finished in a rough wrinkle gold/bronze usually faded to a darker color.
H=19.5"   With this cone it is considered "Rare"

American Bosch Ambotone

Table model with single free edge cone. Stamped metal back with pot metal stand. Frame finished in copper/bronze. Woodgrain micro-veneer cone.
H=18.5" $175-200

American Bosch Ambotone

Table model with single free edge cone. Stamped metal back and pot metal stand. Woodgrain micro-veneer cone. Wicker front cone guard. Finished in copper/bronze or olive green with gold highlighting around base and on wicker front.
H=18.5" $500-600

American Bosch Ambotone

These pictures are variations in design of the Ambotone speaker back. Variety #1 has a brass reproducer mounting with pot metal "T" handle. Variety #2 is the Concert Ambotone with a pot metal reproducer mounting and loop handle. Variety #2 also has a mechanical volume control.

American Bosch Junior Ambotone
Table model with black wood frame and stand. Single cone with blue/black designs in center.
H=12.5″ $80-100

American Bosch Library Ambotone
Table model with gold wood frame and stand. Cone is black with gold dragon design.
H=12.5″ $125-150

American Bosch
Table model with wood cabinet. Grille front with solid back.
H=Approx. 12″ $75-80

American Bosch
Table model with wood cabinet. Grille front with solid back.
H=Approx. 12″ $75-80

Atwater Kent Model E

Table model with single free edge cone showing color variations. All have metal stand and case with crinkle finish. Black and green with smooth metal finish was also available.
H=18" $70-80

Atwater Kent Model E

Floor stand model with gold cast iron and pot metal floor stand. Marbled cone optional.
H=56"   "Rare"

Atwater Kent Model E2

Table model with single free edge cone. Finish is black and green with smooth metal base and frame. Other colors were also available.
H=15.5" $90-100

Atwater Kent E3

Table model with single free edge cone and gold grille cloth. Metal base and frame.
H=13" $80-100

Atwater Kent Model K
Table model with single free edge cone. Metal frame and stand.
H=Approx. 15" $350-400

Atwater Kent F4A
Table model with brown metal base and stand. Electro-dynamic type.
H=Approx. 14" $50-60

Atwater Kent Model F4 and F4A
Table model with black metal base and stand. Electro-dynamic type.
H=Approx. 14" $70-75

Other Atwater Kent speakers not shown are the F2, F2A, F6, F6A, F7, F7A. All speakers are the electro-dynamic type for use on radio models such as the 55F and 55P (25 cycle), 60, 61 and 67.

Front

Back

Arborcone

Table model with metal base and masonite backing. Impedance switch for adjusting speaker input. Original color and design shown.
H=18" $90-100

Ansonia

Table model with wood case, grille front and solid back.
H=Approx. 12.5" $60-70

Exclusive Designs in Unexcelled Speakers

Each of the speakers illustrated embodies desirable principles of Radio reception not to be found elsewhere. Each reflects a new note of charm and beauty and dignity that find a warm welcome in any home.

By turning a small knob the Sylfan "Frusto" can be quickly and easily adjusted and tone quality modulated to produce clear distinct tones and to increase volume. The adjustable device is simple and effective and cannot get out of order. Connection of unit with cone is such as to minimize distortion and to enable the speaker to take larger loads with finer tonal results and without producing second harmonics. There is an utter absence of blasting and chattering from power tubes.

Sylfan "Frusto" has a conical cone—an exclusive Baker-Smith feature—sharply responsive to sound vibrations on a wide range of frequency. It brings out the overtones with full resonance and faithfully reproduces, without distortion, the deep bass notes. Models "G" and "H" are regular cone type speakers, antique finish.

Ask your radio dealer for a demonstration in your home.

Baker-Smith Co., Inc.
Manufacturers and National Distributors
Chronicle Building San Francisco, California

Sylfan "Frusto" Model "F"
$32.50 West of Rockies
$35.00 East of Rockies

Enchanter Model "H"
$32.50 West of Rockies
$35.00 East of Rockies

Enchanter Model "G"
$25.00 West of Rockies
$27.50 East of Rockies

Citizens Radio Call Book , Sept. 1927

Baker-Smith Enchanter Model G

Table model with cast aluminum frame and stamped metal base. The Enchanter nameplate on back of stand.

H=15.5" $325-350

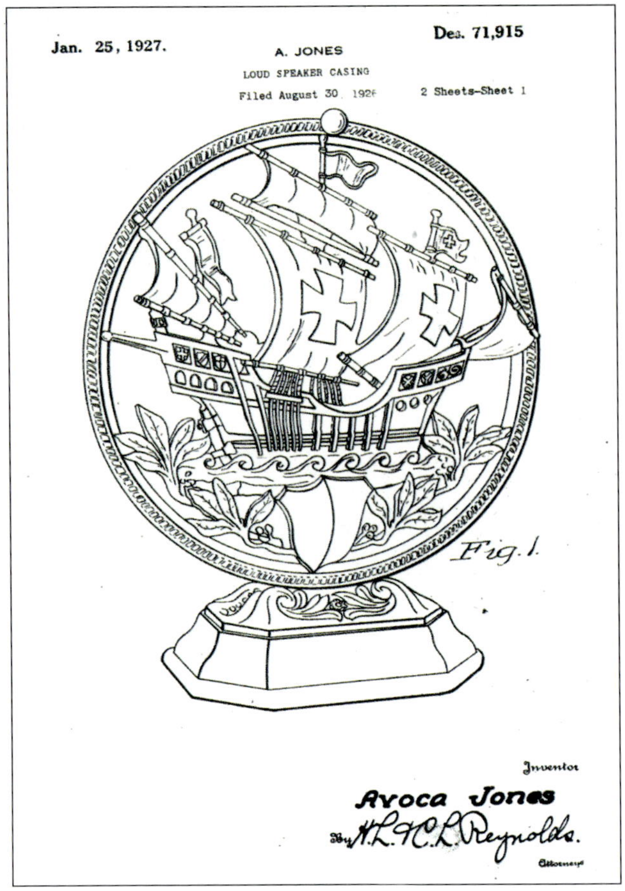

United States Patent Office

Baldwin

Table model with wood cabinet.

H=Approx. 12" $40-50

Brandes

Table model with metal frame and single free edge cone design. Pot metal base with pressed metal housing. Brandes name cast into base front. Internal stand is pot metal. Appears to be home brew but has been documented to be original.

H=16" $125-140

Brandes, Federal Brandes Inc.

Table model with wood cabinet and fleur-de-lis grille design.Gold Brandes name on cone.

H=11" $80-100

Bremer Tulley

Table model with wood cabinet and front grille. Nameplate on front.

H=Approx. 13" $70-80

Brunswick

Table model with wood cabinet. Grille front and back.

H=12.5" $70-75

Cannonball

Table model with single free edge cone. Cast iron metal base. Blue marbled or black crinkle cone with gold highlighting.
H=17″ $90-100

Cannonball

Wall or table model with single free edge cone and wood frame.
H=Approx.16″ $125-150

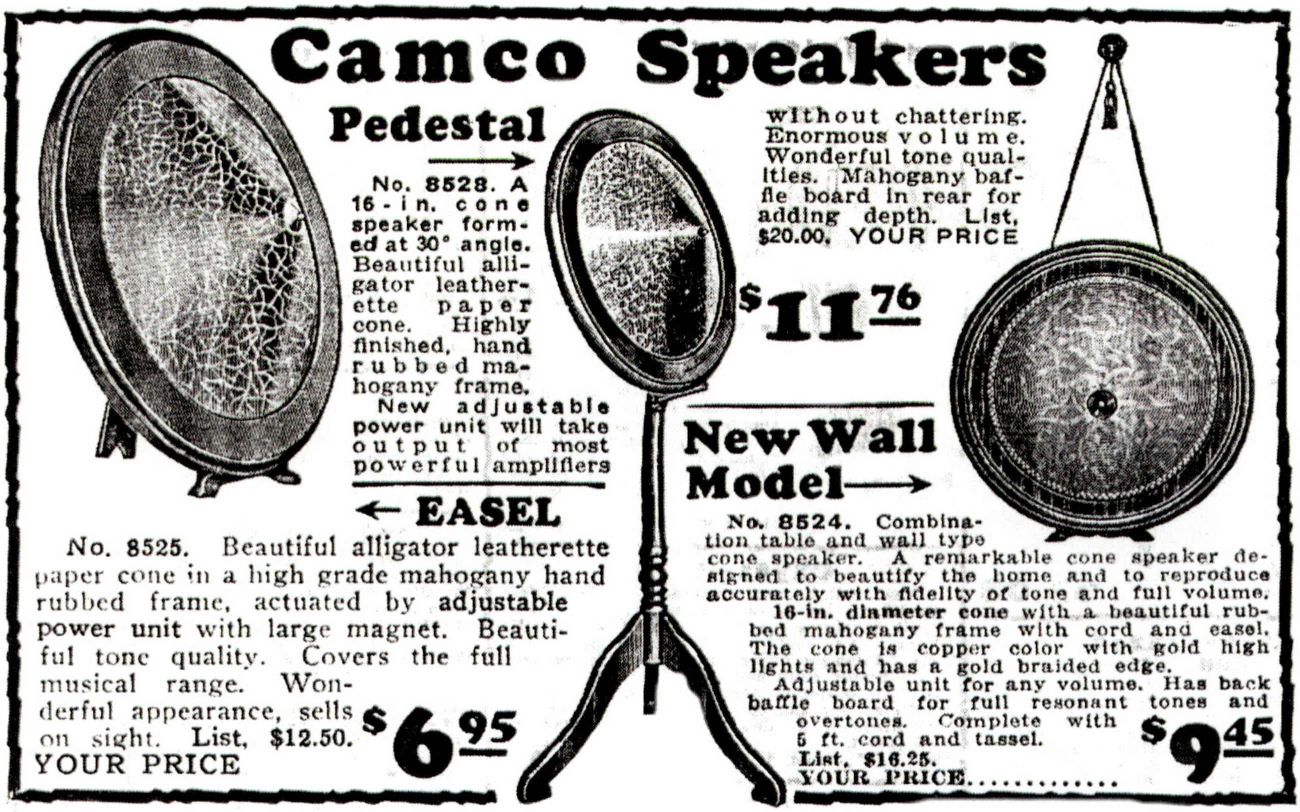

Haskell Radio Corp. Catalog

Claravox Inc. Model S-3
Table model horn speaker design modified for cone speaker.
H=Approx. 19" $175-200

Modified Cone Reproducer

***Radio Retailing*, February, 1927**

The illustrated Model S-3 reproducer is made by Claravox, Inc., Canton, Ohio. The cone or diaphram, which has been treated to protect it from atmospheric conditions, rests on the convex of the standard where it is thoroughly protected by a scross grill. The speaker is also equipped with a special filter system placed in the base of the standard. Two models are being made at present, the S-3 and R-3. Model R-3 has a silk covered grill with a gilt molding set into the opening of the bell. Model S-3 has a polished walnut rim 16½-in. in diam. by 1⅜-in. wide with a black inner and outer edge which, together with a gold inner rim, give it a very attractive appearance. The intended retail price of Model S-3 is $48, that of Model R-3, $42.50.

Radio Retailing, Feb. 1927

Radio, July 1927

Crosley Super-Musicone
Table model with single free edge cone. Metal frame and base finished in smooth bronze/gold.
H=18″ $150-175

Crosley Super-Musicone
Table model with single free edge cone. Metal frame and base finished in smooth bronze/gold. Front and back shown. Cone design is same as used on the smaller Musicone. This cone has been documented but is not common.
H=18″ $175-180

Crosley Type E
Table model with metal base and frame. Crosley crystal brown finish. Electro-dynamic with dual coils for push-pull output.
H=13" $90-100

Crosley
Table model with pot metal frame. Sunburst grille cloth design.
H=Approx. 11" $40-50

Crosley Ultra-Musicone
Table model with Greek Key design front and back. Crosley crystal brown finish.
H=14" $90-100

Crosley Musicone Deluxe
Table model with wood cabinet. Same internal design as the small Musicone without base.
H=14" $80-90

Crosley
Floor console with type E speaker. Blue conewith gold highlighting.
H=Approx. 27" $125-150

Crosley Musicone
Floor stand with metal lyre center design. Model E speaker mounted on back.
H=Approx. 30"   "Rare"

Improved Musicones

Musicones improve the reception of any radio set. They are perfect affinities in beauty and reproductive effectiveness for Crosley Radios. A tilt-table model with brown mahogany finish stands 36 inches high, $27.50—16-inch Super-Musicone as pictured above with "Bandbox," $12.75—12-inch Ultra-Musicone, $9.75.

Radio, October 1927

Crosley Musicone

Table model with single free edge cone. Metal frame and base finished in smooth bronze/gold. Pot metal base.

H=14" $80-100

Crosley Musicone Model D

Table model with metal base and frame. Orange or brown cone. Found in conventional two or three lead magnetic and four lead electro-dynamic type.

H=13" $75-80

DeForest
Shown with F-5 receiver.
Size not available "Rare"

DeForest Audalion
Table model with pot metal housing and cylindrical cone design. Cone was originally a textured brown.
H=18.5" Dia.=10" "Very Rare"

DeForest
Radiophone Type W-6 " Renaissance". Picture taken from Radio Retailing Nov. 1925
H=Unkown "Rare"

DeForest Model LS-400

Table model with metal stand and pot metal reproducing unit. Cone design light blue tint with brown background usually faded to a darker color. Holes around cone are for flexibility.

H=Approx. 14" $125-150

DeForest Model LS-400

Same as above except with different cone design.

H=Approx. 14" $100-125

DeForest Model LS-400

Internal speaker. Normally found in the F-5 portable. Special Cone material.

H=Approx. 13"   Normally not sold separately

Dictogrand Roll Speaker
"Standard" table model. Cone consists of one large roll for lower notes and a smaller upper roll for higher notes.
H=12" W=16" D=6" $175-200

Dictogrand Roll Speaker
"Deluxe" table model. Cone consists of one large roll for lower notes and a smaller upper roll for higher notes.
H=12" W=16" D-6" $225-250

Radio News, Nov. 1926

Dixie Reproducer Model A

Table model with wood mirror design.
H=Approx. 14" "Rare"

Duval

Table model with cast iron round frame and ship.
H=Approx. 14" "Rare"

DUVAL AIRPLANE

Type C Complete with unit. Reg. $15 $10.85

Famous for its reproduction. It employs two diaphragms.

Duval Ship Model Cone Speaker

Reg. $12.50 Beautiful to behold and listen to. $5.95

All new Magnetic and Dynamic Speakers in Stock, such as Jensen, Magnavox, R. C. A., Peerless. Amplion, New-combe-Hawley and others

New York Sun, Sept 1927

Duval

Table model with cast iron octagon frame and ship.
H=Approx. 14" "Rare"

Economy Products
Table model with very unusual and ornate case. Additional information not available.

Empire Electrical Products Co. Model 102
Table model with double free edge woodgrain cone. Heavy cast iron base and stand with mechanical volume control on back.
H=16.5" $225-250

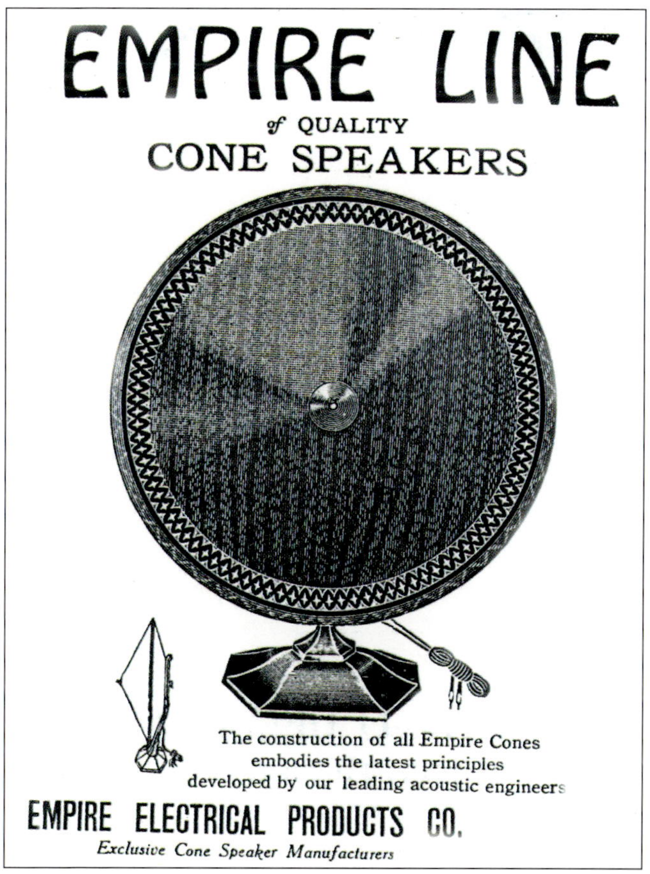

Ensco

THE ORIGINAL AND WORLD'S FINEST 3 FOOT SPEAKER KIT

And Now a Complete Line of "ENSCO" Models

The famous "Ensco" three foot cone has made radio history. Now we are introducing four new "Ensco" models:—two of them just like the old reliable 3 foot models but in the two foot size. And two with a beautiful polychrome pedestal, one three foot and one two foot. The new models round out the "Ensco" line and make it by far the most attractive on the market.

COMPLETE KITS

The "Ensco" Kits are complete. Every necessary part is included together with a twelve page illustrated instruction book which explains in detail just how to make six styles and three sizes of cone and roll speakers. The "Ensco" can be assembled in less than an hour.

ABSOLUTELY GUARANTEED

We guarantee the "Ensco" to be the equal of any manufactured speaker regardless of price. The "Ensco" produces the truest tones you have ever heard. After you hear your set with the "Ensco" Cone, you will wonder how you were ever satisfied with your former speaker.

AT YOUR DEALER OR DIRECT FROM US

Go to your dealer, ask him to demonstrate the "Ensco"—compare it with any speaker regardless of price. If your dealer has not been supplied send the coupon to our nearest office. Shipping charges paid. Our absolute money-back guarantee protects you.

See them at the Chicago Show

STANDARD KIT $10.00

WALL KIT $11.00

OTHER MODELS

No. 10-24 Standard Kit in 24" size $10.00
No. 11-24 Wall Type Kit in 24" size $11.00
No. F-135-24 24" Cone with special polychrome pedestal $13.50
No. F-145-36 36" Cone with special polychrome pedestal $14.50

SEND THIS

Radio News, November 1927

Ensco

This is one of the largest cone speakers manufactured in the 1920's. It could be purchased in kit form or pre-assembled. The picture shown is an original 36" diameter cone. Speaker back was made of wood with an easel stand or rope and tassel for wall hanging.

Diameter 36"   "Rare"

Eveready Model 2 National Carbon Company

Table model with octagon metal case. Olive green finish with silver stripes. Eveready grille with cloth on front and dust cover on back.

H=9.5" W=9.5" $140-150

Eveready Model 1 National Carbon Company

Table model with wood cabinet and grille.

H=11" $80-90

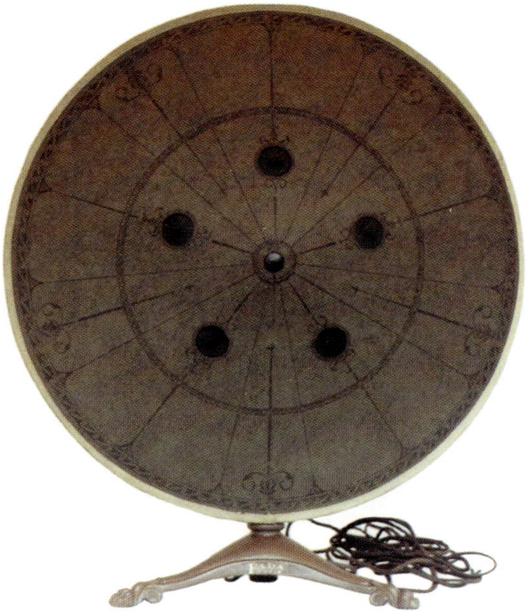

Front

Back

FADA

Original brochure shown. Very few are known to exist due to deterioration of the pot metal frame. FADA name across the middle of base.

"Rare"

(A)

(B)

Farrand

Table model with metal stand and frame. Single free edge marbled or woodgrain cone.
(A) Model 24 Concert Senior oval design. (B) Model 20 Junior oval design.

(A) H=26" $175-200
(B) H=20" $125-150

Farrand Senior

Table model with metal base. Round double free edge cone with gold rope around edge. Same cone design front and back. Cone also found in woodgrain.

H=20.5" $150-175

Farrand Junior

Table model with metal base. Round double free edge cone with gold rope around edge. Same cone design front and back. Cone also found in woodgrain.

H=16" $125-150

Farrand Model 60

Table model with Gothic design. Wood case with ornate pot metal front grille.

H=Approx. 12" $75-100

Farrand

Brass speaker tag used on the metal frame oval speakers.

Farrand Model 30-P

Table model Concert Grand with slant wood frame.

H=20" $125-150

Farrand Model 62

Panel speaker. Can be wall or cabinet mounted. Dust cover on back.

H=Approx. 14" $60-70

Freed Eisemann Model 147

Table model with single free edge cone with or without cone spring retainers. Design found in gold or black. Pot metal base with stamped metal stand.

H=18" $175-200

Freed Eisemann Model 345

Table model with wood cabinet shown in walnut. Also found in silver, green or black.

H=12" $80-100

Freed Eisemann

Copy of original instruction manual supplied with Model 147 speaker.

Fine Arts, Plaza Music Co. Logo

Fine Arts Plaza Music Co.

Table model with single free edge aqua and gold fabric cone. Cast iron base and frame.

H=18" $500-550

Fine Arts (Magnet)

Table model with cast metal base and frame. This speaker was Manufactured by Plaza Music Co. but may have a "Magnet" or Fine Arts nameplate.

H=18" $250-275

Radio Retailing, Nov. 1926

Front View

The FITCH

Back View

The unit is adjustable. It will not chatter or blast.

This, together with the use of two cones of unequal size and angle, gives the FITCH its remarkable clarity and fidelity of tone.

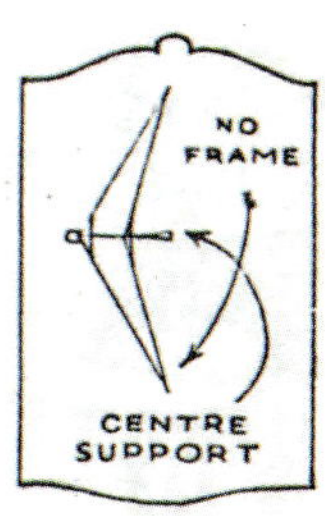

$12.50

FITCH RADIO COMPANY

105 West 40th St. ~ New York City

Radio News, Nov. 1925

Fitch Radio Company

Table model with metal stand and double free edge cone. Unique double convex cone as can be seen in the ad. Floral pattern around front section of cone. Mechanical volume control on back.

H=15" "Rare"

Grebe

Table model with double free edge cone. Pot metal base and stand.

"Very Rare"

National Geographic

Front

Back

Herald

Table model with cast aluminum metal base and stand. Mechanical volume control on back.

H=18″ $150-175

Herald Brass speaker nameplate

Radio News, Oct. 1928

Haskell Radio Corp. Catalog

Jewett Cone

Table model with metal base and paper frame. Gold silk front cover with metal in center.Reproducer is located in the base with a Pyralin or papier-maché air column. Opening of air column is toward the back. Sound is reflected toward the front from the cone shaped back.

H=17.25" $175-200

Jewett Cone brochure.

Kolster K6

Table model with repwood grille front and back. Wood case with metal claw feet. Colors found in walnut, cream and antique green. Manufactured by Brandes for Kolster.
H=12" W=12" $90-100

Allied, 1930

Lansing

Table model with metal base and wood sounding board. Grille cloth on front with metal speaker cover on back.
H=Approx. 14"   Rare"

Leaf Burkhard

Table model with cast metal base and wood frame. Double free edge cone. Also designed for wall hanging with rope and tassel.
H=22" $130-140

Leaf Burkhard

Table model with cast metal base and stand.Wood cone mounting ring. Same design as the Walcone except the Leaf Burkhard name on back of cone.
H= Approx. 20" $125-135

Magnavox Cornell
Table model with metal case and wood base. Also a variety exists with grille cloth on front and slightly different logo on back.
H=10" $70-75

Magnavox Stanford
Table model with wood mantel clock case design.
H=10.5" $70-75

Magnavox Warwick Model 1103
Table model with wood sounding board and metal base.
H=Approx. 15" $80-90

Radio News, December 1926

Magnavox logo

Metro Cone
Table model with single free edge 17" cone. Cast iron base with aluminum frame and mechanical volume control.
H=24" $450-500

Metro Cone
Table model with double free edge 22" cone. Cast iron base and mechanical volume control on back.
H=24" $450-500

Metro Cone
Floor stand model with single free edge cone. Cast iron stand with aluminum frame and mechanical volume control
H=64" $600-650

Metro Cone
Floor stand model with double free edge cone. Cast iron stand with mechanical volume control.
H=64" $550-600

Mid-West Per Vois
Table model with woodcase and very heavy crackle finish. Gold grillecloth.
H=Approx. 15" "Rare"

Mid-West Per Vois
Speaker nameplate on back of speaker.

Natural Tone

Majestic Cone

Designed for use with or without power tubes

Reproduces music or voice *exactly as it comes from the broadcasting station.* A full, rich, *pure tone,* free from distortion.

A beautiful cone, 16 in. *in diameter,* ruggedly constructed, laminated wood frame and metal base, with sensitive, adjustable unit, finished in rich mahogany. Packed individually in fibre air-cushion cartons. Standard package of 6.

PRICE, EACH $2.75

Write for Other Special Bargains Now to

SPIEGEL RADIO COMPANY

135 Liberty Street New York City, U. S. A.

TERMS: 2% Discount Allowed on C.O.D.
3% Discount If Cash Is Sent With Order

Radio News, August 1928

SHIPS OF MELODY

LA PINTA

A picturesque Merchant Ship of the Fifteenth Century.

Size, Height 25 inches.
Width 13 inches.
Length 27 inches.

SANTA MARIA

THE FLAG SHIP OF COLUMBUS IN THE DISCOVERY OF AMERICA IN 1492.

Size: Height, 25 in. Width, 11 in. Length, 27 in.

MAYFLOWER

THE SHIP THAT BROUGHT THE PILGRIMS TO AMERICA.

Size: Height, 25 in. Width, 9 in. Length, 27 in.

BUILD A BEAUTIFUL SHIP MODEL LOUDSPEAKER---*in a few hours*

No tool needed but a tackhammer—Any schoolboy can put it together.

PERFECT TONE - PLENTY OF VOLUME - NO DISTORTION

Do you want a Loud Speaker that is different and worth $100.00?

Do you want a beautiful Ship Model?

Do you want perfect tone? Of course you do. All of these are to be found in the Melody Ship, and it will not cost you as much for both, combined in one, as the ordinary Ship Model or Loud Speaker would alone. The Ship Model when complete is the most beautiful of its kind on the market today. The Loud Speaker combination is unsurpassed for beauty of tone and undistorted volume that brings out all the low musical notes faithfully reproducing the zoom of the base viol, the drums and tuba, as well as the higher alto and soprano. There is nothing like it on the market today anywhere. It is absolutely new. This is not an experiment, but a real proven fact, sold on a money-back guarantee to assure satisfaction.

THE WORLD'S LARGEST BUILDERS OF SHIP MODELS will supply you with all the parts to build a Melody Ship. You may choose either the La Pinta, Santa Maria or Mayflower Models. The parts for the Ship Models are made of wood. This is not a cheap cardboard imitation, but a real Ship Model. Putting this Model together is as simple as A, B, C. The instructions read like this: put No. 49 on top of No. 48 and tap lightly with a small hammer, etc. A schoolboy can put it together in a few hours.

The Loud-Speaker Unit is of the Electro-Magnet type. Power amplification is not needed to force the low tones through. They come through with perfect ease and do not interfere with the higher notes. Giving you faithful reproduction at all frequencies. The Unit is placed on the Main Mast, which is firmly seated two inches deep in a three and one-half pound hull. Making it impossible for counter vibrations to effect the perfect reproduction of the Melody Sail. The driving pin is attached to our super-vibrating, especially prepared, Melody Sail. It does not change the appearance of the Ship Model. Instructions for installing come with each Melody Ship.

MINIATURE SHIP MODELS
3818-20-22-24 Baring Street
Philadelphia, Pa.

Those who own a Ship Model and wish to install the Melody Sail and Unit can easily do so by forwarding us an out-line drawing of your main sail. We will then forward to you a Melody Sail and unit to fit your Model, via parcel post C. O. D., nine dollars ($9.00), plus postage.

The un-assembled parts for either the La Pinta, Santa Maria or Mayflower Models (without the Melody Sail attachment) can be had for four dollars and ninety-eight cents ($4.98) plus postage. Full detailed diagram and instruction sheet, showing and explaining every operation, for completing the Model supplied with each. If you want the un-assembled parts for either the Santa Maria or Mayflower Model, together with the Melody Sail attachment, for making the Melody Ship complete, send us your order—NO MONEY—pay postman $12.50, plus postage.

CANADIAN AND FOREIGN ORDERS MUST BE ACCOMPANIED BY MONEY ORDER

If, after assembling the Melody Ship, you do not think it worth many times the purchase price, return it to us in good condition and we will gladly refund your Money. What can be fairer than this?

WRITE FOR OUR CATALOGUE

Complete Melody Ship Orders Include Metal Figures Illustrated Above. If Figures are Ordered Separately Same Will Be Furnished For 25c Each Figure.

MINIATURE SHIP MODELS — S.I. July
DEPARTMENT L16.
3818-20-22-24 Baring St., Philadelphia, Penna.

Please send me the complete parts cut to fit and ready to put together for the MELODY SHIP Complete..................For which I agree to pay postman $12.50, plus postage. For the SHIP MODEL ....only. For which I agree to pay $4.98, plus postage. For MELODY SAIL & UNIT only for ship model...............for which I agree to pay postman $9.00, plus postage.

PLEASE PRINT NAME AND ADDRESS

NAME ....
STREET OR R. F. D. ....
CITY OR TOWN ....
STATE ....

Science and Invention, July 1927

Miniature Ship Models, Inc. Philadelphia, Pa 1927

The Merchantman Melody Ship speaker is one of a series of ship speakers that were manufactured by this company. They could be purchased as complete kits or retrofitted to an existing ship. Factory assembled ship speakers could also be purchased. The hull and masks are constructed from solid wood. Cast iron sailors were optional and could be purchased for fifty cents each. A total of sixteen figures were available . The ship pictured has eight figures, and was factory assembled. Paint and all parts are original as shipped from the factory. It also is a lighted version with illuminated port holes and three treasure chests located on the ships deck. The main sail serves as the speaker cone with reproducer unit attached to the main mask.
H=27" L=33" W=14" "Very Rare"

Miniature Ship Models, Inc. La Pinta

Original colors. This may be a Melody ship retrofitted with a speaker kit.

See ad for size "Rare"

Miniature Ship Models, Inc. Venetian La Pinta

Shown in original colors. The Santa Maria ship is very similar in size, color and design.

See ad for size "Rare"

RADIO'S
Greatest Triumph

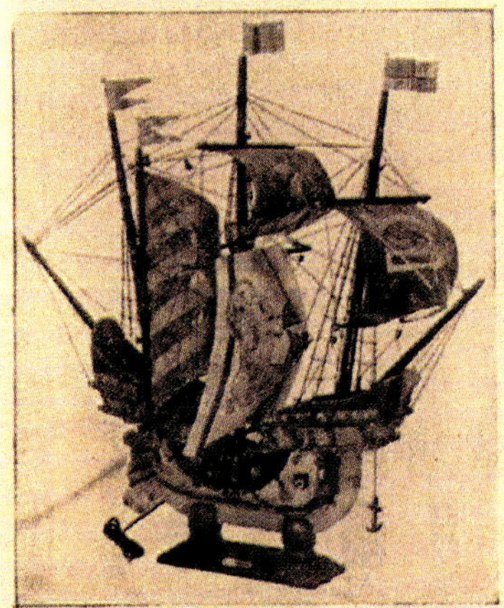

THE MELODY SHIP

Size: 26" high; 12" wide; 27" long (overall)

You can put these ship model loud speakers together in a few hours of pleasant pastime

You need not know anything about ship building or loud speakers to assemble one of these models. As a matter of fact if you have never seen a ship you can put the ship model loud speaker together in a few hours by following our diagrams and instruction sheet which is so written that a school boy can understand it. All you need is a small hammer. The parts are numbered on the diagram to make the process even more simple. As long as you can understand such instructions as these (actually taken from assembling chart that is included with every model) you cannot fail: "Take No. 57, place it in front end of No. 56 and tap lightly with hammer. Next take No. 58 and place it up against No. 57 and tap with hammer to bring it into place."

Easy? Nothing simpler. The instructions are just like that from the beginning to end. Do this and that — and before you realize it a beautiful model has grown before your eyes.

The MELODY SHIP is made entirely of wood. A positive free edge cone speaker with a unit of the electro magnet type, powerful enough to operate a 72-inch cone, is attached to the main-mast. This insures splendid service and ample volume with no distortion.

The loud speaker is made in the shape of a sail and does not alter or change the appearance of the ship in any way. The sails are artistically painted with the various designs and figures of the period represented by the models.

The tone of this loud speaker is far superior to any cone of its size on the market today. This is made possible by our patented melody sail.

Write for our beautiful illustrated catalog. Fill in the coupon below and we will act upon it immediately.

MINIATURE SHIP MODELS, Inc.
3818-20-22-24 Baring St.
Philadelphia, Pa.
Canadian Branch:
1485 Bleury Street, Montreal, Canada

MINIATURE SHIP MODELS, Inc., Dept. Y
3818-20-22-24 Baring St., Phila., Pa.
Please send me the complete parts, for the loud speaker ship model cut to fit, and ready to assemble for the La Pinta as shown above. I will pay postman $12.50 plus postage (a few cents).
PLEASE PRINT NAME AND ADDRESS PLAINLY

Name..
Street or R. F. D. ....
City..
State...

Radio, October 1927

Mohawk

Table model with wood case and and front grille cloth. Solid back with green Mohawk nameplate.

H=16" $90-100

Mohawk Deluxe Pyramonic

Table model withwood case and mahogany finish. Mohawk nameplate on back.

H=Approx. 17" $125-150

Music Master ad.

Mohawk

Metal nameplate pinned to back of speaker cabinet.

Music Master Model 12

Table model with wood base and stand. Mahogany finish. Pleated gold silk with rosette and rope trim on front. Back is gold silk with rope trim around outer edge and gold trim around center cone adjustment access. Aluminum nameplate with Music Master logo on top.

H=16.5" W=18 $175-200

Na-Ald
Octagon laminated pressed paper back.
H=4" $100-125 with original box.

Na-Ald
Round laminated pressed paper back.
H=4" $125-150 with original box.

The Na-Ald Midget Cone Reproducer

Description: **This is an unusual cone reproducer, chiefly in the matter of size. It is designed to fill a unique purpose and is not just a "trick." Its tiny volume is in proportion to its size; it is intended for use at times when great volume of reproduction is not desirable—such, for instance, as late at night when the family have gone to bed and the DX fan wants to listen in without waking the household. Another practical scheme is to use several of these units in different rooms. Their cost is low and a number of them may be installed for the price of one standard-sized reproducer. A specially made small unit is attached to a rod that drives the miniature paper cone. The edge of the cone is bound with felt and rests against the composition sounding board. The small composition base is finished to match the octagonal sounding board. It is a standard cone in all respects but size.**

Maker: **Alden Mfg. Co.**

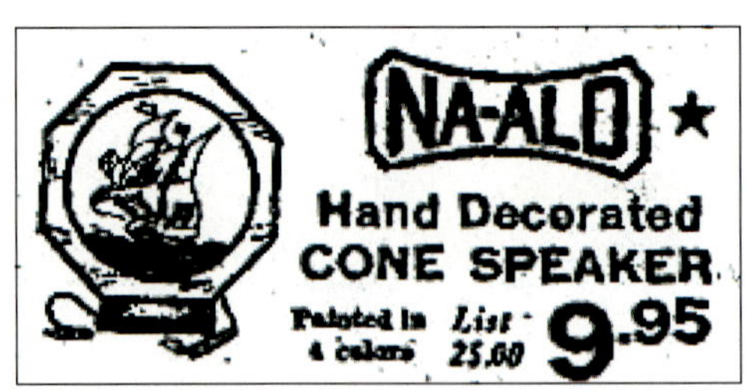

Na-Ald
Table model larger than miniature speakers. Wood base and sounding board. Painted paper cone with ship design.
H=16", W=14.5" New York Sun Sept. 1927 "Rare"

Neutrowond
Table model with wood cathedral case finished in American walnut. Removable back with nameplate.
H=12" $85-90

Newcombe Hawley Model 83
Table model with wood case and grille.
H=11" $30-40

Newcombe-Hawley
Table model with wood case and front grille.
H=11" $30-40

Newcombe-Hawley
Table model with cast iron base and frame. Sea shells on feet. Single free edge cone. Has small "NH" stamped on front of main sail.
H=12.5 "Rare"

O’NEIL DYNAMIC SPEAKERS

Pictured above is one of the best all-round values obtainable in a dynamic speaker. The best buy from the standpoint of performance —the best buy from the standpoint of appearance—the best buy from the standpoint of sturdy construction—and finally, the best quality speaker purchase you can make.

The fact that it is built by O'Neil is worth more to the purchaser than all the specifications that could be written into print. The cabinet is of miniature proportion compared with the average dynamic speaker cabinet. It has been made as small as it is practical to build such a cabinet. It is finished in our regular two-tone Radio Walnut color to match the best radio sets.

Model 507. For use with 6 volt storage battery or 6 volt battery eliminator.
List Price **$45.00**

Net Price, $27.00

Model A 51. With A. C. Dynamic Unit. Same cabinet as used for the D. C. model but equipped with A. C. Dynamic Unit dry disc rectifier, powerful transformer, etc., ready for operation by simply plugging into your light socket. List Price **$50.00**

Net Price, $30.00

Model D 508. Either 110 volt or 220 volt D. C. Dynamic Unit. This model is only used with current supplied from D. C. lighting circuit. List Price **$45.00**

Net Price, $27.00

O'NEIL DYNAMIC UNIT

D C Model D 509. For operation with (6 volt) battery or battery eliminator.
List Price **$30.00**
Net Price, $18.00

D C Model D 511. For D. C. light socket operation, 110 or 220 volt. List Price, **$30.00**

Net Price, $18.00

A C Model A 510. For A. C. light socket operation, equipped with dry disc rectifier and powerful transformer 110 volt, 60 cycles.
List Price **$40.00**

Net Price, $24.00

O'NEIL DYNAMIC TABLE

This beautiful walnut consolette is equipped with the new Dynamic Unit, so much desired by those requiring volume, with unusual fidelity of tone and reproductive qualities. The Unit functions directly from any convenient 110 volt A. C. 60 cycle house current supply. Size: Depth, 17 in.; width, 26 in.; height, 30 in. List Price **$67.50**

Net Price, $40.50

O'NEIL STANDARD CONE

This new model is finished in beautiful mahogany bronze, and the cone is tan with a mottled effect in green. It is equipped with our new No. VIII Unit and Cone Assembly.
List Price **$14.75**

Net Price, $8.85

O'NEIL MAGNETIC UNIT CHASSIS

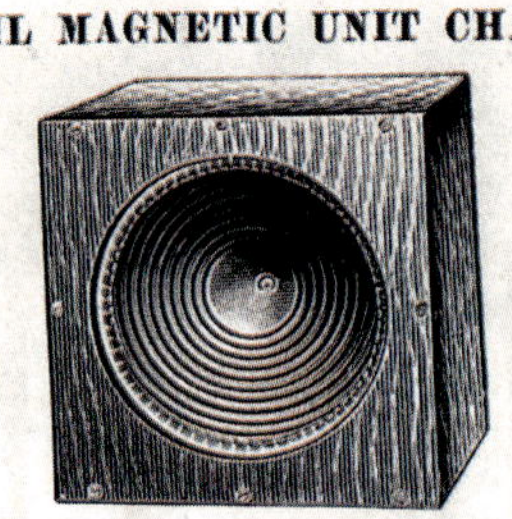

Model C B 18. This model is designed to fit the standard type cabinet, drawer, etc. Furnished with Baffle Box as illustrated. Complete with 5 ft. telephone cord.

Licensed under Lektophone Patents

List Price **$18.00**

Net Price, $10.80

O'NEIL CLOCK MODEL

Designed to match the finer radio set—its beautiful lines and exquisite Radio Two-Tone Walnut finish will appeal to the most critical eye. Equipped with the Super X Magnetic Unit. The elusive bass notes, uncommon in speakers having a small cone, are all in evidence here, reproducing the quality of the argest and best cones.

Model C 105. List Price **$30.00**

Net Price, $18.00

O'NEIL ILLUMINATED SHIP

Design licensed under U. S. Patent No. 71915

A beautiful generous sized oval frame full-rigged ship constructed of cast aluminum and finished in Antique Bronze—illuminated with two concealed 25 watt red and blue intermediate lamps, effecting a beautiful sunset and glow illumination. Every home has several attractive lamps; here is a speaker and Lamp in one, which will beautify any room.

Model L S 19. Equipped with Lamps, Plug, Lighting and Phone Cords. Stands 20 in. high; weight, 11 lbs. List Price **$25.00**

Net Price, $15.00

Our Special Service Division is at your command.
WE CAN FURNISH
Any Receivers, Kits, Parts, Speakers or Electrical Accessories Not Shown In This Catalogue.
Write for Special Quotations.

ROYAL-EASTERN ELECTRICAL SUPPLY COMPANY, 16 WEST 22ND STREET, NEW YORK

The Royal Eastern Radio Catalogue

O'Neil Model 8508

Table model. Antique green and gold highlighting on wood frame with decorative plaster design. Metal base with Griffins.
H=17.5" $350-400

O'Neil Model D-99

Table model. Antique bronze and gold highlighting on wood frame with decorative plaster design. Metal base with Griffins. Also found with the name American Automotive.
H=17.5" $350-400

O'Neil Junior

Table model with oval cast iron ship. Found in gold, or bronze.
H=16" $300-350

O'Neil Model LS-19

Table model with oval cast iron ship. Illuminated version with two 120 vac lights behind ship.
H=20" $500-550

O'Neil

Table model with metal base and wood trim frame. Cone design found in gold or black.

H=16" $100-110

O'Neil

Table model with heavy metal base and wood frame. Gold cone design.

H=16" $100-110

O'Neil

Table model with wood base and frame. Gold cone design.

H=14" $80-90

O'Neil

Table model with metal base and metal encased wood frame on back. Gold cone design.

H=14" $90-95

Operetta

Table model with metal case and front wood grille. Very similar to Victor and Westingale.

H=Approx. 14" $125-150

Hamilton-Carr Corporation Catalog

Old Kentucky Home

Table model with cast iron base and stand. Mechanical volume control on back. Double free edge cone. Cone design has not been documented. Distributed by Belknap, Louisville, Kentucky.
H=Approx. 20"   "Rare"

Ortho Cone

Table model with cast iron base and stand. Double free edge cone with ornate design on front. Mechanical volume control on back.
H=64"   **$400-500**

Ortho Cone

Floor stand model with cast iron base and stand. Double free edge cone with ornate design on front. Mechanical volume control on back.
H=64"   **$475-500**

Pacent Cabinet Cone

Table model with wood cabinet and Moorish design. Also found with mahogany finish.
H=11" "Rare"

Type A Pacent Cone, 17 inches in diameter.
Manufactured under Lektophone Patents.

The
Immediate Success
of the
PACENT CONE

THIS remarkable new development of the Pacent engineers reproduces the delicate inflections of the human voice and musical programs with such startling truth and clearness that it has met with *immediate success* among radio users.

Its tone range is practically unlimited, and it can be used with any receiving set, with or without power amplification. A superior sound reproducing device from the standpoint of absolute fidelity, crystal clarity and above all, musical range.

All types are equipped with a suitable length cord and Pacent detachable plug.

Type A, table type, 17 inches, with handsome bronze base. List price .... $28.50 ($31.50, West of the Rockies)

Type SA, 36 inches, mounted on American Walnut stand for floor use. List Price ...... $89.50 (Slightly higher West of the Rockies)

Type WA, similar to Type SA, but designed for mounting on wall.

If your dealer cannot supply you, write us for interesting and informing literature.

PACENT
Radio Corporation
156 West 16th Street

Radio News, Dec. 1926

Pacent Senior

Table model with metal base and stand. Double free edge cone with stick and rope design front and back. Gold rope around edge and back cover. Also found with micro-veneer woodgrain cone.
H=21" $175-200

Pacent Junior

Table model with metal base and stand. Double free edge cone with stick and rope design front and back.
H=Approx. 18" $150-175

Pal Butterfly
Table model with cast aluminum base and frame finished in gold. Brown cone with gold highlighting. Also supplied with optional floor stand.
H=Approx. 14" "Rare"

Pal Liberty
Table model with cast iron base and frame. Also supplied with optional floor stand. Nameplate on back of base.
H=18.5" $1000-1200

Note: The Pal Manufacturing Company later became the American Reproducer Corporation.

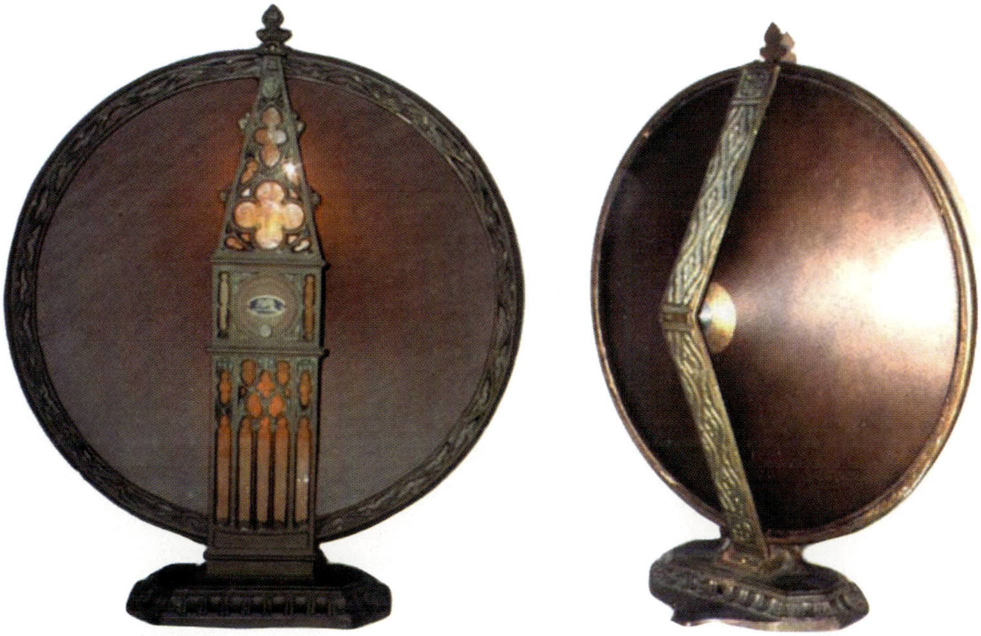

Pathe'

Illuminated table model showing front and back. Back view has the cone protector assembly and anti-tipping base. Two versions of the cone mounting have been documented. The above unit has a curved stamped metal rim with a spring encased in a fabric sleeve that holds the cone. The second version has a cast iron rim and beveled edge for cone support. Finish was originally gold/bronze but has faded to a much darker color. Cone is brown with gold design.

H=22" $450-500

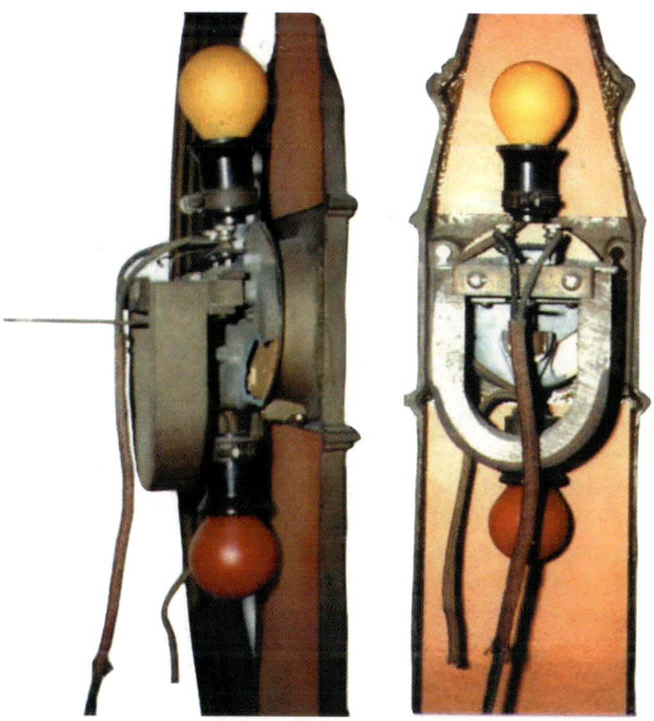

Pathe'

Illuminated version showing side and back view of the 120 vac light assembly. Bulbs and socket assembly are original.

Pathe' ad. *New York Sun* Sept. 1927

Pathe'

Table model cathedral design with cast iron base. Found in three frame types. 1. Cast iron taper fit cone mount. 2. Stamped metal rim for cone mounting. 3. Cast iron removable taper fit flange cone mounting. Finish is gold or old gold.
H=22" $300-350

Pathe' Grecian (Caduceus)

Table model with medical symbol.Cast aluminum or cast iron stand and frame. Cone mounting is either tapered flange or flange screw clamp. Plain tan/brown cone with optional light blue/green marbling. Larger version has been found with cone protector. Also known as the Pathe' Caduceus.
H=22" $450-500

Barwicks Radio Buyers Guide 1927-28

Pathe' Speaker Nameplate

This brass tag is usually mounted on the front side of the Pathe' speaker and covers the mounting screw of the reproducing unit. It is made of brass with black background and red rooster. Sizes vary.

Pathe'

Table model with metal stand and frame. Found in nickel plate or mahogany finished frame. This was one of the first cone speakers marketed for general public use. Mechanical volume control on back.

H=14.5"   "Rare"

Pathe' Grecian (Caduceus)

Floor stand model with very ornate cast iron base and twisted brass stand. Also known as the Pathe' Caduceus.

H=56"   "Rare"

The Wireless Age, Feb. 1924

Radio Call Book, 1929

Front

Back

Peerless

Table model with wood case and grille front. Removable back with nameplate.

H=12" $80-90

Peerless

Table model with metal case and open front. Peerless name on cone.

H=10" $30-40

Pennsylvania Speaker

Table model with cast iron base and stand. Music notes cast into base. Single free edge cone with mechanical volume control. Cone shown is not original. Original cone design has not been documented. Distributed by Hamburg Brothers Pittsburgh, Pa.

H=16.5" $150-175

Haskell Radio Corp. catalog

Philco
Table model with cast metal case. Front tapestry grille.
H=Approx. 12" $50-75

Philco
Table model with stamped metal case. Grille front with solid back.
H=10.75" $40-60

Philco Cathedral
Table model with wood case and volume control on front. Open back with large heavy magnet.
H=Approx. 12.75" $125-150

Pausin Engineering Octacone

Table model with metal case and spacer posts between base and frame.
H=12" $90-100

Pausin Engineering Octacone

Floor stand model with ornate cast iron base and twisted metal stand. Attached cast floral designs.
H=Approx. 48" $450-500

Pausin Engineering Octacone

Table model with metal case without spacer posts between base and frame.
H=12" $90-100

Quam
Table model with metal case. Black and gold crinkle finish. Metal grille cloth retaining ring.
H=10" $55-60

Quam
Table model with metal case and gold/bronze crinkle finish. Metal grille cloth retaining ring.
H=Approx. 13.5" $65-70

QUAM Model "A"

Just as there is a difference in the tone qualities of the human voice, so are there great differences in the speaking voices of Radio. That speaker approaches nearest to perfection which has the widest range in accurate tone reproduction. It must pick up and reproduce highest tones of the violin and deepest tones of the pipe organ. For all of these—select the Quam. The outstanding feature of the Quam Loud Speaker is its unmatched range—its ability to give you ALL. Finished in rich antique brown, it will grace the finest Radio Set by its distinctive beauty, and thrill the most critical audience by its remarkable reproduction of over-tones. The most music-sensitive ear will appreciate its faithful reproduction of the entire musical scale. You will marvel at its absence of "blasting." It is a beautiful, lasting, marvelous "New Voice for Radio."

No. 2K-01209. Retail Price, $15.00.
Your NET $9.80

RCA Model 103
Table model with octagon repwood frame and tapestry front. Cone and cloth dust cover over back.
H=14.75 **$175-200**

RCA Model 100
Table model with metal case. Bronze/gold grille clothe front and back.
H=15" **$65-70**

RCA Model 103
Same as table model with lighted floor stand.
H=46" **"Rare"**

RCA Model 100
Same as table model with metal floor stand.
H=46.5" **$140-150**

RCA Model 100A

Table model with pot metal housing. Grille cloth front and backwith cardboard bottom cover and label.
H=10" $60-80

RCA Model 106 Floor console

Tapestry front and both sides with closed back. Electro-dynamic type requiring 120 vac for operation. Power switch located on side.
H=Approx. 30" $175-200

RCA Model 100B

Table model with pot metal front and back. Woodgrain sheet metal body. Found with silver or gold grille cloth design.
H=8.5" $70-75

R. C. A. Model 104 Loudspeaker

Radio, July 1927.

RCA Model 100A

Same as table model except with cast metal floor stand. RCA stamping in base.
H=45.5" $300-350

R.F.I.
Table model with ornate oval metal case and cast iron base. Mechanical volume control on front.
H=20", Width = 20" $400-450

Radio News, Dec. 1926

R.F.I. Junior
Table model with double free edge cone and gold cone design. Metal base and frame. Cone is brown with gold design.
H=17 $130-140

R.F.I. Senior
Floor model with cast metal base and stand. Double free edge cone with brown design.
H=55″ $350-400

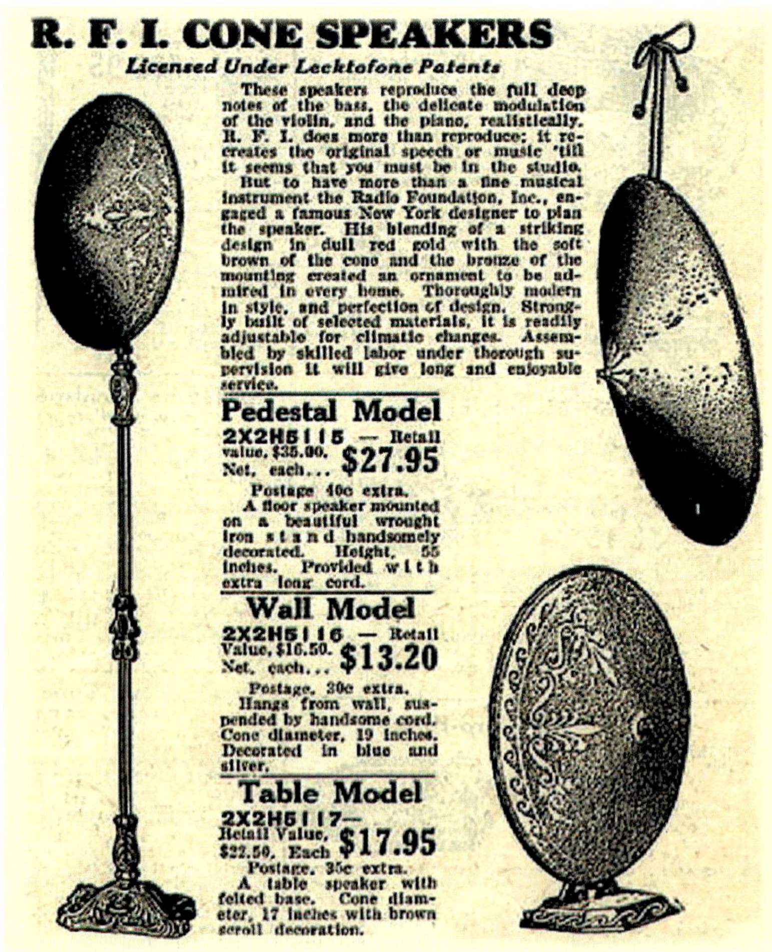

Radio Magazine

Rola Model 106 Pedestal Recreator
Wood floor stand and sounding board. Metal speaker housing on back. See back view of table model.
H=36″ $150-175

Rola
Table model with metal base and wood sounding board. Metal speaker housing on back.
H=16″ $140-160

Rola
Table model with wood cathedral case.
H=Approx. 12″ $50-60

Rola
Table model with tall wood cathedral case.
H=Approx. 15″ $60-70

Radio, Dec. 1926

SAAL

Table model with metal stand and base. Unique double free edge cone with offset apex. These speakers were also marketed by Sears Roebuck as the WLS Silvertone.

H=17" "Rare"

Safar "Victoria"

Table model with solid bronze figure and wood base. Mechanical volume control, on front, disguised as a bugle. Italy 1927
H=Approx. 22" "Rare"

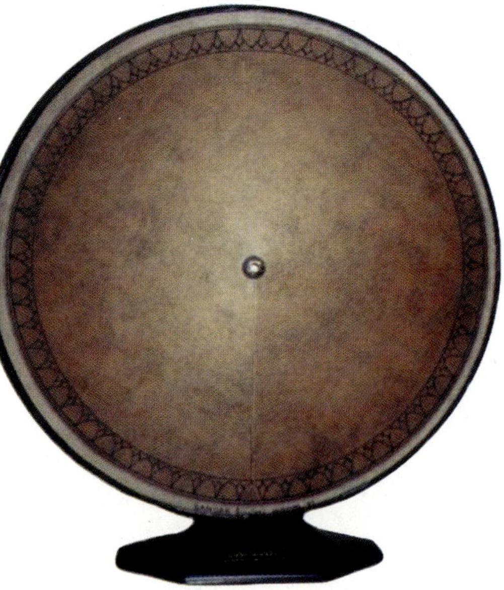

Sandar
Metal table model with single free edge cone. Found with black or gold cone designs.
H=16″ $110-120

Sandar
Table model with micro-veneer wood grain cone. Double free edge with gold rope around edge.
H=24″ $150-175

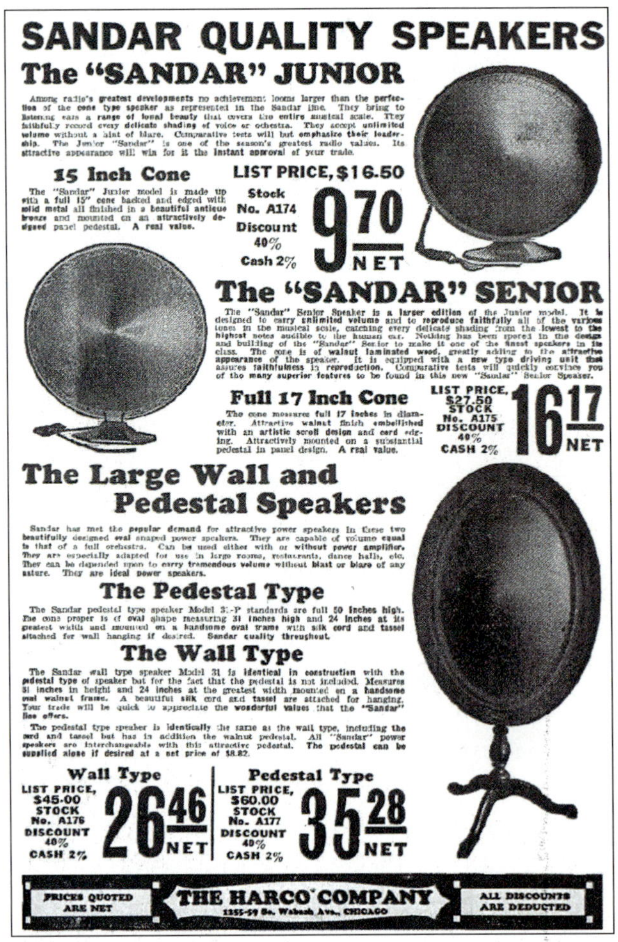

The Harco Company

Selector Co. Seattle, Washington

Table model with cast metal base and frame. Very ornate with gold cone designs. This speaker usually is displayed backward due to its beautiful eagle design on back.
H=Approx. 17"   "Rare"

Senesco

Table model with double free edge cone. Cone design may not be original.
H=Approx. 18" $150-175

Slagle

Table model with metal base and wood sounding board. Gold fabric cone with tilt top for wall mounting.
H=16.5" $130-150

Silvertone WLS

Table model with unique offset apex cone design. Metal base and stand with lever type volume control on back. Double free edge cone with fabric edge and gold floral arrangement in center. Manufactured by SAAL and distributed by Sears Roebuck.

H=17" $175-200

Silvertone Cone Speaker $11.75

Unquestionably one of the greatest values of the season in radio. Big reproducers of similar quality often sell for $20.00 to $25.00, yet we ask only $11.75. Equipped with a large, heavy magnet which has a sensitive, floating type armature that assures the utmost in clarity and realism throughout the entire musical scale. Built to handle the higher voltage of power tubes and is guaranteed to handle volume without any sacrifice of true tone quality. Made of a heavy russet brown parchment in both front and back; neatly decorated and reinforced around outer edge of cone. Mounted on a strong metal frame and pedestal; bronze finish. Diameter, 19 inches. Height, over all, 21¼ inches. Shipping weight, 23 pounds.

57K2241¼.....$11.75

Smaller cone, same as above in both design and construction. Diameter, 14 inches. Height, over all, 16½ inches. Shpg. wt., 13 lbs.

57K2240¼...........$7.95

Sears Roebuck Catalog, 1927

Silvertone WLS
Table model with metal base and stand. Double free edge cone with fabric edge and gold diamond pattern in center.
H=17" $130-135

Silvertone WLS
Table model with metal stand and base. Single free edge cone with fabric edge. Very similar to Herald.
H=18" $150-175

Silvertone WLS
Table model with metal case and grille.
H=Aprox. 13" $50-60

Silvertone WLS
Table model with wood case and gold trim on grille insert.
H=Approx. 14" $70-100

Radio News, November 1927

(A)

(A)

(A)

(B)

Sonochorde Senior

These speakers were manufactured in various designs with a single free edge wood grain cone and mechanical volume control through front of grille clothe. The octagon design, type A, is more common and is found in various colors. The metal frames were usually mahogany in color but faded to a darker black color. They were also found in smooth old gold or rough crinkle light gold finish. The mahogany style has either a wine or bronze colored silk . The gold type has a bronze silk only. Type B, has a cast aluminum round mahogany frame with gold crinkle highlighting. Wine color silk has also been found for type B. The round model has an open cone, unlike the octagon model with metal cone protector. Type B is very delicate and easily broken. Cones also have a woodgrain finish. Type A manufactured in Chelsea, Mass. Type B manufactured in Revere, Mass.

Type A H=18" $200-225 Type B H=18" "Rare"

Sonochorde Junior J-28

Table model with single free edge and woodgain cone. Mahogany finished metal frame with wine or bronze silk.

H=14.5" $140-150

Sonochorde

Senior, Junior and Round back view. Woodgrain pie cone design. Metal housing over cone is the same for Senior and Junior. The round model does not have a metal cover.

Sonochorde Speaker tag

This warning tag was attached to the speaker cord.

Sonochorde Senior Model A-5

Floor stand model with octagon cone. Same design as the table model. May have also been supplied with bronze silk.

H=55" $300-500

Sparton

Table model with cast metal base and stand. Base has three claw feet. Double free edge cone with Greek key design around front edge and center spears. Speaker is identical to the Akradyne Model 20 except center cone design. Smaller versions of this design have been found but not confirmed as a Sparton.
H=26" $275-300

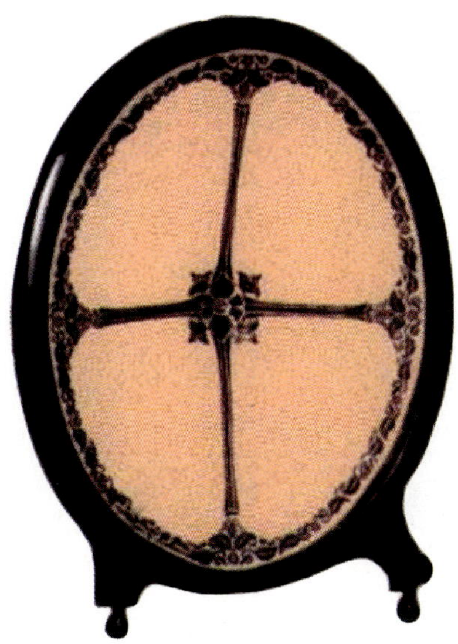

Sturrup-Plimpton-Newell Co.

Floor or table model with wood frame.
H=44",W=30.5", D=11.5"   "Rare"

Splitdorf ConeTone Art Model

Wood back with papier-maché stand and cone support. Very ornate with elliptical fabric cone. ConeTone label on back.
H=26", W=22" $450-500

Splitdorf

Table model with walnut detachable sounding board for wall hanging. Rope and tassel are gold. Gold fabric cone. Cone usually faded to an olive green.
H=20" $150-175

Splitdorf

Walnut wall hanging model. Same as the table model without base.
Dia. 16" $140-150

Stevens

Table model with wood Gothic design. Open grille.
H=Approx. 11" $50-60

Stevens

Table model with wood base. The famous Stevens "Burtex" fabric cone. This material was manufactured by Stevens and was supplied on many other speakers such as the Splitdorf, Algonquin, Symphonic Globe, etc. Laminated from different fabrics, it was known for its moisture resistance. Cone was originally gold but usually faded to an olive green.
H=Approx. 16.5" $80-90

Stevens Speakers The Golden Voice of Radio

Stevens Speakers reproduce tones just as they come from the entertainers. Clear and accurate of the entire scale. Write.

STEVENS MFG. CORP., 46 East Houston St., New York

Radio News, October 1928

STEVENS SPEAKERS

Model CT 27

A full 17-inch Burtex conoidal diaphragm. Equipped with the exclusive Stevens armature unit with two point suspension and specially designed output transformer. Frame finished in hand rubbed Jasper brown mahogany. Burtex diaphragm finished in subdued gold bronze. Height over all 22¼ in. This is the identical model as illustrated. Supplied with elegant picture cord for hanging on wall. No. 8505. List $35.00. YOUR PRICE $20.58

Model B 27

No. 8506. A specially designed 17-inch Burtex diaphragm. Similarly finished and attractively hand rubbed. Stands 21¼ in. high. Similar in appearance to model illustrated. List $25.00. YOUR PRICE $14.70

Model A 27

No. 8507. A full 14½-inch Burtex seamless diaphragm built on the famous Stevens conoidal principle. List $18.50. YOUR PRICE $10.88

The frame and base is similar but not identical in appearance to the type illustrated.

Candelabra Model ET 27

Beautiful pierced standard, 5½ feet high. Graceful and filigreed, with relief designs. Encircled by an elegant Jasper brown mahogany finished frame hand rubbed to beautiful sheen. Its full 19-inch Burtex diaphragm is finished in gold bronze. Actuated by the specially designed Stevens armature unit—built in output transformer. Used with or without power amplifiers of high voltage. Equipped with 20 ft. extension cord. No. 8503. Retail, $65.00. YOUR PRICE $38.22

Hanging Wall

The Stevens Burtex conoidal hanging wall model, 22 in. in diameter, finished in hand rubbed Jasper brown mahogany. May be used with power amplifier without injury to windings. Furnished with a beautiful silk picture cord with hanging rosette. No. 8504. List $45.00. YOUR PRICE .. $26.46

Haskell Radio Catalog

Stewart-Warner

Table model with metal case and bronze/gold crinkle finish. Floral design around front edge.

H=7" $60-70

Stewart-Warner Model 435

Table model with metal base and frame. Speaker body is pot metal with pressed metal rings for grille support.

H=12" $50-55

Stewart-Warner Model 425

Table model with dancing girls on front. Pot metal base and frame. Finish is found in antique bronze or with bright brass. The dancing girl design is also found in floor console models. Brass design is not as common as the antique bronze and should be valued slightly higher.

H=13" $275-300

Stewart-Warner reproducers employ the new Williamson single armature magnetically balanced unit which is designed to give equal reproduction of all tones from the lowest double-bass to the highest

Stewart-Warner Model 420. *Stewart-Warner Model 425.*

piccolo. Model 420 has a drum 9 in. in diameter and 5 in. deep. It gives great volume with minimum power consumption. It is metal finished in bronze golden brown and has silk fabric mesh front and back. Model 425 has a drum diameter of 11¼ in. and width of 6¼ in. Its finish is like that of Model 420 with the addition of a graceful silver silhouette superimposed on the front mesh.

Radio, July 1927. Silver version has not been documented.

Stromberg-Carlson 5A

Floor stand model with wood stand and sounding board. Found in two versions. Version #1 shown with metal sounding board support and cast iron anti tilt weights on feet. Version #2 has wood sounding board supports with a round cast iron anti tilt weight attached to center post. Version #2 has slightly different stand design. Model 7 is the wall hanging version of this speaker.

H=35" Floor stand model $350-400 Model 7 is considered "Rare"

QST May 1926

Stromberg-Carlson

Speaker filter with optional jack adapter. Can be used with any Stromberg or most other speakers of this era.

Stromberg-Carlson Model 10 & 16
Floor Stand model with metal rim. Both are identical in appearance with slightly different reproducer units.
H=35" $350-400

Stromberg-Carlson Model 11, 17 & 19
Wall Hanging unit. Same as the model 10 &16 except with wall hanger bracket including rope and tassel.
Dia.=22" $250-300

Stromberg-Carlson Model 14 & 15
These speakers appear to be identical except for a filter located internally of the model 15. Pot metal frame and base with grille cutouts in back.
H=17" $125-150

Symphonic Globe Speaker

Table model with cast iron stand and fabric cone. Old world globe speaker housing with cutouts on bottom of globe. Top of globe tilts open to expose speaker cone.

H=Approx. 20" with 12" globe. "Rare"

The Symphonic Globe radio loudspeaker is a library globe whose exterior is an up-to-date map of the world and whose interior is a

Symphonic Globe Radio Loudspeaker

tone-chamber for a loud speaker unit. It is claimed to give remarkable fidelity of reproduction. The globe can be cleaned with a damp cloth.

Radio, July 1927

Symphonic Free Edge Cone

A $9.50 Retail Value for $3.85

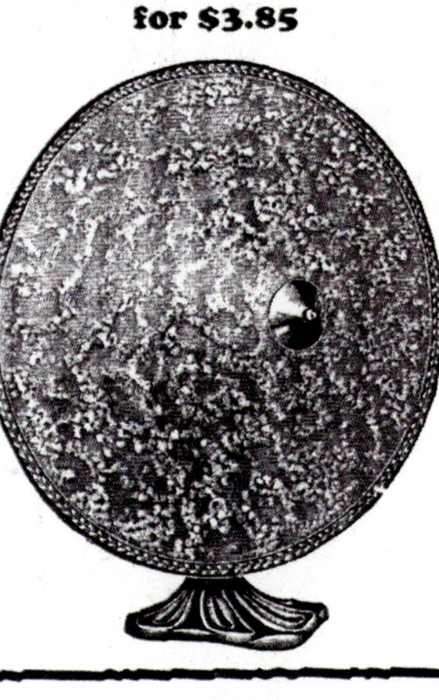

Marvelous Tone Quality

A REMARKABLE new 16-inch free edge cone speaker. Scientifically designed to cover the lowest and highest frequencies of the musical and audible scale. The new sensitive unit with its oversized heavy magnet responds to the slightest impulse. It will faithfully reproduce the barely audible low frequencies of the pipe organ or the base viol. The sturdy, heavy, adjustable unit will not rattle or chatter at the higher frequencies. It is almost impossible to overload this new unit.

No. 8529

Retail List $9.50

YOUR PRICE $3.85

Beautiful Design

The cone diaphragm is made of beautiful alligator leatherette parchment paper. Its coloring and design is attractive and its appearance will harmonize with the settings of the finest furnished home. The braid edge of the cone gives it a rich finishing touch. The metal stand with an artistic leaf design is finshed in a soft harmonious shade to blend with the color scheme of the cone.

Tremendous Volume

The new power unit of the Symphonic Cone is built to deliver enormous volume and is adjustable to the volume suitable for any type receiving set and any home.

Remarkable Low Price

The Symphonic Free Edge Cone is priced to you unbelievably low. Its regular list price is $9.50 but on a basis of comparison with most speakers it is well worth $18.00. We have chosen the Symphonic Cone as a leader in its field. Its beauty and its efficiency will sell it on sight for profits in excess of what is possible with most lines of merchandise.

New 20-Inch Cone

Large superb free edge cone speaker. Diameter, 21 inches. Beautifully finished. Wonderful tone quality. Deep natural tone.

No. 8530. Special value.

List Price $25.00

YOUR SPECIAL PRICE........ $5.95

Haskell Radio Corp. catalog

Teletone
Wood table model with mechanical volume control on front. Mahogany finish.
H=14" $80-90

Haskell Radio Corp. catalog

Teletone
Wood floor stand model with tilt top. Cone is not original design.
H=Approx. 26" $125-150

Thermatrol Manufacturing Co.
Midget cone speaker with metal base and back. Unique marbled front cone.
H=Approx. 4.5" "Rare"

Thermatrol Manufacturing Co.
Original midget speaker box.

Thorola speaker label

Thorola Model #9

Table model with molded repwood base and metal frame. Open cone with decorative wood front grille panel. Brown and black marbled cone. Gold Thorola logo on bottom of cone.

H=17" $185-200

Thorola Model #18

Table model with molded repwood base and metal frame. Back has metal screen cone guard with gille cloth. Decorative wood front panel with gold Thorola logo at bottom.

H=17" $185-200

Thorola

Table model with wood telephone style stand design. Swivel reproducer unit with single free edge cone. Gold trim with wire edge cone support.

H=17" "Rare"

Front

Back

Timmons Concert Grand

Very ornate cast iron frame and base with oval shaped paper cone. Unique offset cone apex. Timmons Concert Grand logo on back of cone. Stand bottom also has a Timmons label.
H=38" $500-550

Side

Barwicks 1927-28

Timmons Talker

Table model with wood mirror design frame. Single free edge cone with Timmons logo on back.

H=16" $150-175

Timmons Talker

Table model with metal case and front grille cloth. Solid metal back.

H=11" $50-60

Radio News, Nov. 1927

Tower Featherweight
Table model with metal base and stand. Single free edge similar to the Tower Scientific versions but smaller base design. Mechanical volume control on back.
H=15" $130-140

Tower Scientific Version I
Table model with cast iron base and stand. Single free edge cone with ornate gold design on cone. Cone design usually faded to a darker color. Mechanical volume control on back.
H=17" $175-200

Tower Scientific Version II
Table model with cast iron base and stand. Single free edge cone with green and brown symbolic fleur-de-lis design on cone. Mechanical volume control on back.
H=17" $175-200

Tower Castle Cone
This speaker was designed after the tower trademark. It can be found with or without light gold marbling on blue/green cone. It is slightly eliptical in design with the frame being wider than tall. It has a unique stand with the name "Tower" cast into the base. Lighted versions have not been documented but may have existed.
H=17.75" $450-500

Tower (no name)
Table model with cast iron frame usually found with the Tower logo on back of cone. Not all cones have the Tower logo.
H=12.25 $275-300

Tower Model 28
Commonly known as the Knight speaker. Metal housing with repwood front, back and base. Finish is a mahogany.
H=15" $135-140

Tower Meistersinger
Wood table model with marbled cone. Also has an attachment on back for wall hanging.
H=18.5" $140-150

Tower Adventurer
Table model with cast iron frame. Found in white with bronze highlighting, gold or bronze. Cones are blue/green with optional marbled cone. Orange Tower logo on back. Variations have been found with the moon and stars painted on the cone but are not documented as being original.
H=18" $375-400
Note: One variety found 17.5" and thought to be a proto-type.

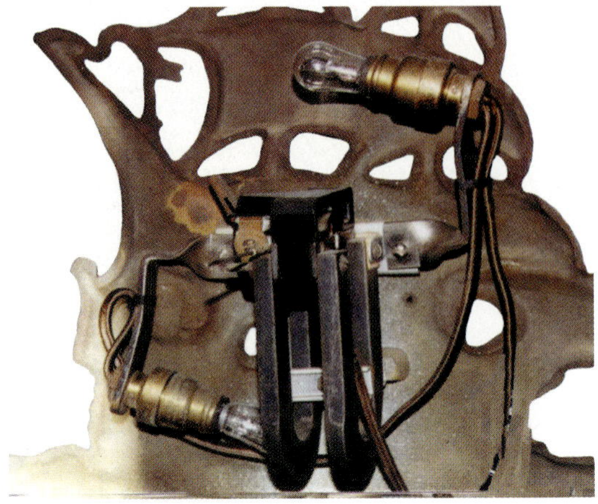

Tower Adventurer
This picture is the optional light assembly on the back of the Adventurer ship frame. Lighted version is not common and considered "Rare".

Tower Pirate Ship
Table model with cast iron frame and blue/green cone. Marbled cones were optional. Found in bronze or off white.
H=14.5" Bronze color $250-275 White color $275-300

Front

Back

Trimm Model 80B

Table model with wood stand and housing. Very ornate carving front and back. Brass nameplate on back.

H=Approx. 20" $400-450

Trimm Model 80

Floor stand model same design as the 80 table model.

H=Approx. 34" $450-475

Trimm Model 11

Table model with metal stand and frame. Single free edge cone design with ornate bronze flower and vine. This model also found without the metal cone protector ring. Both models have a mechanical volume control on back. Cone attachment for the model having a cone protector is a wire and screw arrangement, where as the version that does not have the cone protector , has a stud protruding from the driver with star and knurled nut assembly. Cast iron bases are identical.

H=19" $175-200

Trimm Junior (Cabinette)

Table model with metal stand and frame. Single free edge cone with gold floral design. Does not have volume control.

H=Approx. 17" $150-175

Trimm Concerto Grand Model 38
Table model with single free edge cone. Cone is brown with bronze floral design. Does not have volume control. Very similar to the Model 11.
H=19″ $175-200

Trimm Concerto Model 28
Table model with single free edge cone. Cone is brown with gold floral design. Does not have volume control. Very similar to the Trimm Junior.
H=Approx. 17″ $150-175

Barwick Catalog 1927-28 Edition.

Walter Rowan Battery Supply Co.

Utah Book Speaker
Table or wall model designed to appear as an open book. Reproducing unit vibrates page centers to produce an audible sound.
W=Approx. 17″ H=Approx. 13″ "Rare"

Radio News, Nov. 1927.

Utah speaker label

Utah
Table model with power unit. Requires 120 vac.
H=Approx. 12″ $125-150

Utah
Table model with cast metal frame and crinkle gold finish.
H=9″ $75-80

Utah
Table model with metal frame and base finished in crinkle gold.
H=11″ $60-65

Utah Model 30

Table model with wood frame and pot metal feet. Decorative pressed wood design with Utah logo on front.
H=17" $275-300

Utah Model 30

Table model with wood frame and base. Decorative pressed wood design with Utah logo on front.
H=17" $275-300

Utah Model 16

Table model with mahogany finished metal housing and repwood base.
H=15" $100-110

Utah

Table model with wood housing and front grille. Closed back.
H=Approx. 10.5" $60-100

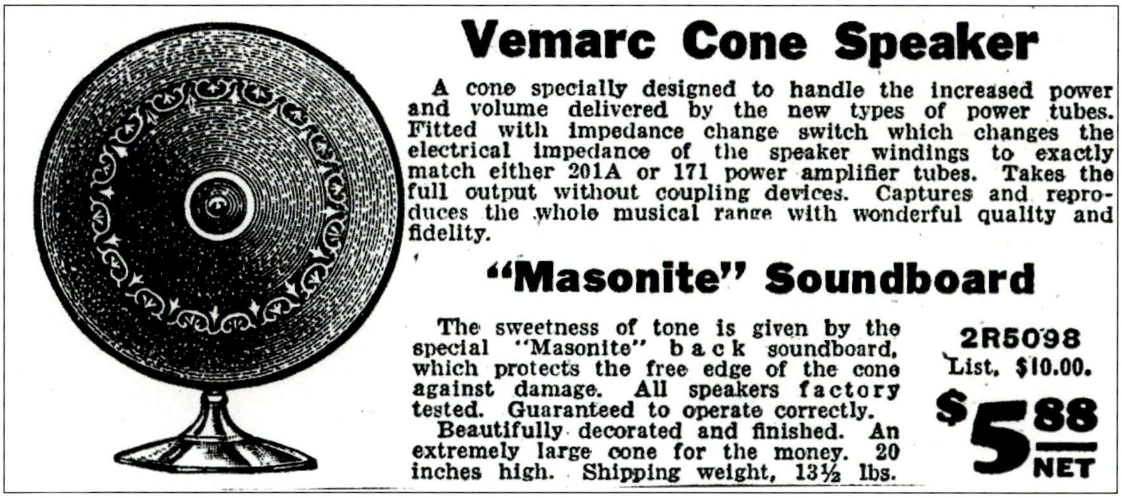

Radio, 1929 Edition

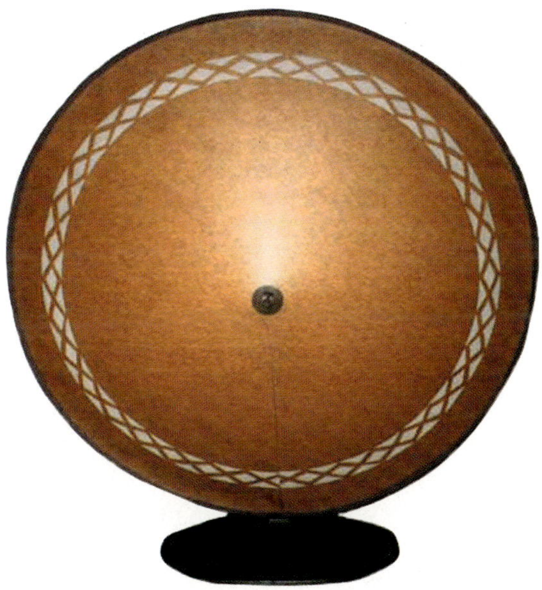

Vitalitone

Table model with double free edge cone. Cast metal base with wood cone support on back. This speaker is uniquely different fron the usual Vitalitone design. It has a paper cone with painted design and fabric cone edge protector. The Vitalitone logo is embossed into the front upper center of the cone and is very difficult to detect. No documentation has been found other than the name on the cone.
H=19" "Rare"

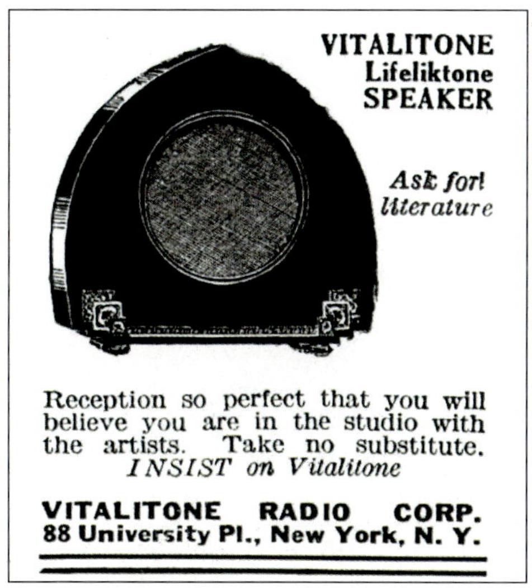

Radio News, Oct. 1927

Vitalitone Ship

Table model with cast iron frame and single free edge cone. Original colors are shown but other speakers have been found with one red cross and one yellow cross. A slightly different variety has been documented with a variation in the reproducer. Cone design is the same diameter but is deeper to accommodate the larger reproducer. Cone designs are blue, brown or blue front and brown back.

H=15.5" $250-300

Vitalitone Ship

Table model with cast iron frame and single free edge cone. Original colors are shown. This model was touted as the "New Vitalitone Ship Model". It is very ornate with sea creatures cast into the frame. Oyster and pearl at top of frame.

H=16.25" "Rare"

Barwicks Co., 1927-28

Radio News, October 1927.,

Victor
Table model with wood housing and front grille.Closed back with Victor label.
H=Approx. 13.5" **$75.85**

Victor Speaker label.

Victor Lumiere Model 1
Table model with folded cones front and back. "His Masters Voice" logo in center of cones. Mechanical volume control on top with "His Masters Voice" nameplate.
H=13.5" **$250-275**

Victor
Table model with metal housing and wood grille. Very similar to Westingale and Operetta.
H=Approx. 14" **$125-135**

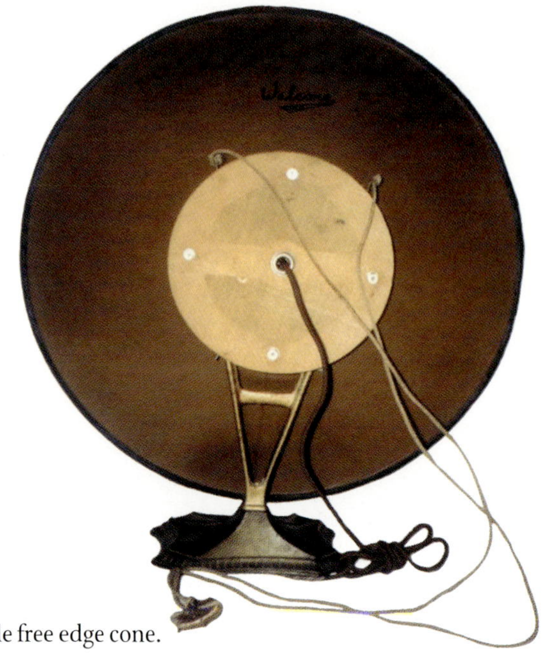

Walcone

Table or wall hanging model with double free edge cone. Rope and tassel for wall hanging

H=approx. 20" $130-140

Walcone

Wall hanging with double free edge cone. Rope and tassel for wall hanging.

Dia.=17 $100-110

Western Electric instruction manual cover

Floor model 42" high

Wall model 36" Dia.

Western Electric Model 548-AW

This speaker is one of the largest manufactured for commercial and home use. It is a 36" diameter and 9" deep double free edge cone. Two wall hanging models were available. Model 548-DW has a very ornate swivel wall bracket. Model 548-CW has a three point wall bracket. All models were 36" diameter with the Greek key design. Floor stand models have been found with ball and claw feet or rounded feet. Stands are identical with some having gold trim.

Floor model $1200-1500,
Wall model with swivel bracket "Rare"
Wall model with three point hanger $800-1000

Western Electric 540-AW

Table model with double free edge 18" diameter cone. Cone design is black diamonds with red hightlighting.

H=21" $250-275

Western Electric 560-AW

Table model with double free edge 24" cone. Cone design is black and red.

H=26.5 $500-550

Western Electric 560-AW

Wall hanging model with 24" double free edge cone. Cone design is same as the table model.
H=Approx. 26.5"   Rare"

Northern Electric Model R-540

Canadian Western Electric Div. Table model with double free edge cone. Brown Cone with black and gold design.
H=21" $325-350

Western Electric Company

Albany
Atlanta
Baltimore
Birmingham
Boston
Brooklyn
Buffalo
Charlotte
Chicago
Cincinnati
Cleveland
Columbus
Dallas
Davenport
Denver
Detroit
Duluth
Grand Rapids
Houston
Indianapolis
Jacksonville
Kansas City
Los Angeles
Memphis
Milwaukee
Minneapolis
Nashville
Newark
New Haven
New Orleans
New York
Norfolk
Oakland
Omaha
Philadelphia
Pittsburgh
Portland
Providence
Richmond
St. Louis
St. Paul
Salt Lake City
San Francisco
Savannah
Seattle
Spokane
Syracuse
Tacoma
Youngstown

A NATIONAL ELECTRICAL SERVICE

International Western Electric Company

195 Broadway **New York, U. S. A.**

Foreign Associated and Allied Companies

AFRICA, SOUTH—
Western Electric Company, Ltd., Johannesburg.

ARGENTINE—
Cia Western Electric Argentina, Buenos Aires.

AUSTRALIA—
Western Electric Company (Australia), Ltd., Sydney.

AUSTRIA—
Vereinigte Telephon und Telegraphen Fabriks A. G., Vienna.

BELGIUM—
Bell Telephone Manufacturing Co., Antwerp.

BRAZIL—
International Western Electric Co., Inc., Rio de Janeiro (Rua dos Ourives 91-1).

CANADA—
Northern Electric Company, Ltd., Montreal

CHINA—
China Electric Company, Ltd., Peking and Shanghai.

FRANCE—
Le Materiel Telephonique, Paris.

GREAT BRITAIN—
Western Electric Company, Ltd., London, W. C. 2.

HOLLAND—
Bell Telephone Manufacturing Co., The Hague.

HUNGARY—
United Incandescent Lamps and Electrical Co., Ltd., Ujpest 4 near Budapest.

ITALY—
Western Electric Italiana, Milan.

JAPAN—
Nippon Electric Company, Ltd., Tokyo.

NORWAY—
Western Electric Norsk Aktieselskap, Christiania.

SPAIN—
Telefonos Bell, S. A. Barcelona (Granvia Layetana 17)

STRAITS SETTLEMENTS—
Western Electric Company, Ltd., Singapore.

SWITZERLAND—
Bell Telephone Manufacturing Co., Berne.

Bargains in Amplifiers and Loud Speakers

THERE is available for sale to Bell System Employees a quantity of loud speaking telephones and amplifiers, at greatly reduced prices. The Hawthorne Merchandise Department has been advised that orders for the apparatus described here should be filled in the order in which they are received. As soon as this surplus stock is disposed of there will be no more of this apparatus available.

Western Electric people desiring to take advantage of this exceptional offer, should place their orders through the Employees Personal Purchase Department, at their location.

No. 548-DW
36″ Cone Type Loud Speaking Telephone for Hinged Wall Mounting
Price $5.00

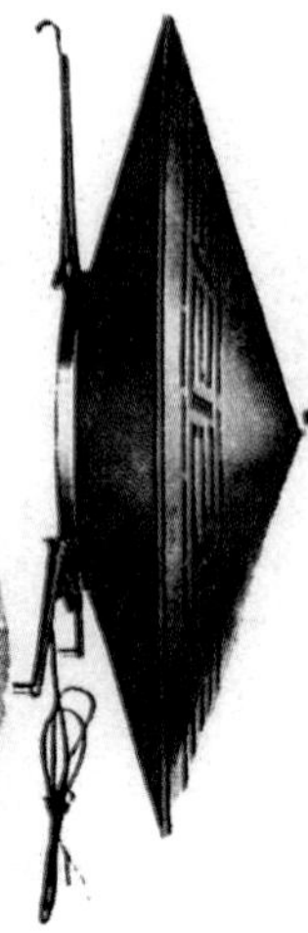

No. 548-CW
36″ Cone Type Loud Speaking Telephone for Wall Mounting
Price $5.00

No. 548-AW
36″ Cone Type Loud Speaking Telephone for Pedestal Mounting
Total height, 49″
Price $5.00

No. 25-B Amplifier
Price $15.00 less tubes

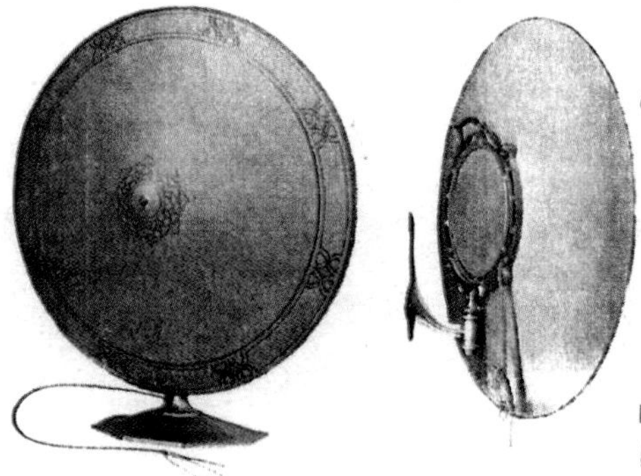

No. 560-AW
Cone Type Loud Speaking Telephone, Diameter 24″; Total Height, 26¼″
Price $10.00

No. 45-A Amplifier
Price $15.00 less tubes

THE *25-B amplifier* is a single stage amplifier which operates directly from 100-120 volt, 50-60 cycle AC supply. It requires no batteries and contains its own transforming and rectifying systems for filament, plate and grid of the amplifying vacuum tube. Input and output impedances are such that it may be connected directly between a radio receiver (nominally 20,000 ohms) and the loud speaking telephones here described. The price of this amplifier is $15.00 less tubes. Two 205-D vacuum tubes are required for operation on the amplifier. The price of the 205-D vacuum tubes is $6.20 each.

THE *45-A amplifier* is the same as the 25-B amplifier except that it also provides a "B" battery supply. The price of this amplifier is $15.00 less tubes. Two 205-D vacuum tubes are required, the price of which is $6.20 each.

THE cone type speakers which are available for use with these amplifiers are the table, wall and floor types. The *560-AW table* speaker is 24″ in diameter and sells for $10.00. There are available three types of 548 (36″ diameter) type loud speakers: *548-AW floor type*, *548-CW wall type* (to be hung from picture molding), and *548-DW wall type* with mounting bracket. The price of the 548 loud speaking telephones is $5.00 each.

Western Electric

This is a copy of the last notifications to the Bell System employees for sale of all existing apparatus. The famous Western Electric loudspeaking telephone and associated surplus equipment was being cleared from inventory never to be available again from the manufacturer.

Westinghouse
Canadian version with wood cathedral shaped housing and grille. Closed back.
H=Approx. 13″ $75-80

Westinghouse
Canadian version with wood cathedral shaped housing and grille. Closed back.
H=Approx. 11″ $70-75

Westingale
Table model with metal housing and wood grille. This speaker is similar to the Victor and Operetta.
H=Approx. 14″ $125-150

Windsor Easel
Table or wall hanging model with ornate cone design. Windsor Furniture Co. name on front of cone.
H=Approx. 22″ $150-175

Radio Broadcast, 1927 and *Radio News*, Feb. 1927

Burns
Pearlescent horn.
"Rare"

Cordonic
Condenser speaker with windmill design.
"Rare"

Exotic Dancer
Table model with single free edge cone. Cast iron base and frame. Cone may not be original color. Manufacturer unknown.
H=17.37 "Rare"

Madera Horn American Art Mache Co.
Die cast wood fibre base and horn. Horn is removable.
H=20" Horn dia. 10" "Rare"

Tower

Midget horn speakers. These horns are made from a very delicate pyralin material. (Left) Color is an opaque brown and red swirl, (Right) Solid black.

H=4" $130-140

Radia Lamp

Table model lamp and horn speaker. Reproducer is mounted in base with sound mirror inside of shade. Lamp shade is not original. Manufactured by the Radio Lamp Co. N.Y.

H=18" $800-1000

Sea Tone

Table model horn with sea shell at top.

"Rare"

Voice from the Sky

Table model horn found in Walnut or Mahogany. Other colors also have been found but not confirmed to be original.

$1200-1500

Ship Speaker

Cast iron ship with handle at top front. Handle has a large "N" cast into it. Unknown manufacturer

Ship Horn speaker

Table model thought to be manufactured by Utah but not confirmed. Horn bell is part of the front deck facing upward.
H=Approx. 27" "Rare"

Yar-Lange "YAR"

Floor model horn speaker. Cast aluminum. Found in bronze or green with gold trim.
H=41" $1200-1500

Kodel

Table model horn speaker with cast horn.
H=8.5" $325-350

Dancing Nymphs
Oval table model with cast iron frame. Thought to be O'Neil but not documented.
"Rare"

M.P.A.
Table model with wood grille. Manufactured on Conduit street in London.

Marvel
Table model with very ornate front grille. Spoke support in back with wire legs. Radio Craft Nov. 1930.
"Rare"

Index

A

Acme Apparatus Co., Cambridge, MA 3,4
Airchrome (Temple) Irvington, NJ & Chicago, IL 5–8
Akradyne Radio Corp., New York City, NY 9, 10
Algonquin Electric Co., Poughkeepsie, NY 11
All American, Chicago, IL 12
American Automotive, Kansas City, KS 57
American Bosch Magneto Corp., Springfield, MA 13–15
Amplion Corp. of America, New York City, NY 12
Ansonia (Radio Foundation Inc.) New York City, NY 18
Arborcone (Also Vemarc) Ann Arobor, MI 18
Atwater Kent Mfg. Co., Philadelphia, PA 16, 17

B

Baker-Smith Co. Inc., San Francisco, CA (Jodra Mfg. Co.) Seattle, WA 19, 20
Baldwin (National Baldwin Inc.) Salt Lake City, UT 20
Brandes (Federal Brandes Inc.) Newark, NJ 21
Bremer Tulley Mfg. Co. Chicago, IL 21
Brunswick-Balke-Collender Co., New York, Chicago & Toronto 21
Burns (American Electric Co.) Chicago, IL 117

C

Camco (Cannon & Miller Co. Inc.) Springwater, NY 22
Cannonball (Cannon & Miller Co. Inc.) Springwater, NY 22
Claravox Inc., Canton, OH 23
Cordonic, Holland, MI 117
Crosley Radio Corp. Cincinnati, OH 24–28

D

Dancing Nymphs (Fictitious name. Unknown manufacturer) 120
DeForest Radio Co., Jersey City, NJ 29,30
Dictogrand (Dictograph Products Corp.) New York City, NY 31
Dixie 32
Duval 32

E

Economy Products 33
Empire Electrical Products Co., New York City, NY 33
Enchanter (See Baker-Smith) 19, 20
Ensco (Engineers Service Co.) New York, Chicago, Boston, Ontario 34
Eveready (National Carbon Co. Inc.) New York City, NY 35
Exotic Dancer (Fictitious name. Unknown manufacturer) 117

F

FADA (F. A. D. Andrea) New York City, NY 36
Fanspeaker Radio Co., New York City, NY & Philadelphia, PA 35
Farrand Mfg. Co. Inc., Long Island City, NY 37, 38
Fine Arts (Plaza Music Co.) New York City, NY 40
Fitch Radio Co., New York City, NY 41
Freed-Eisemann Radio Corp., Brooklyn, NY 39

G

Grebe (A.H. Grebe Co. Inc.) New York City, NY 42

H

Herald Electric Co., New York City, NY 43

J

Jewett Cone (Jewett Radio and Phonograph Co.) Pontiac, MI 44
Jodra Mfg. Co. (See Baker-Smith) 19, 20

K

Knight 43
Kodel Radio Corp., Cincinnati, Ohio 119
Kolster Radio Corp. (Federal Brandes, Inc.) Newark, NJ 45

L

Lansing 46
Leaf-Burkhard Radio Corp., New York City, NY 46

M

M.P.A., Conduit Street, London 120
Madera Horn (American Arts Mache Co) Chicago, IL 117
Magnavox Co., Oakland, CA 47
Majestic Cone (Spiegel Radio Co.) New York City, NY 49
Magnet (See Fine Arts) 40
Marvel 120
Metro Cone (Metro Electric Co.) Chicago, IL 48
Mid-West Per Vois , Chicago, IL 49
Miniature Ship Models Co., Philadelphia, PA 50–52
Mohawk Corporation of IL, Chicago, IL 53
Music Master Corp., Philadelphia, PA & Kitchener, Ontario 53

N

Na-Ald (Alden Mfg. Co.) Springfield, MA 54
Neutrowond (United Radio Corp.) Rochester, NY 55
Newcombe-Hawley Inc., Saint Charles , IL 55
Northern Electric, Montreal, Canada 113

O

Old Kentucky Home (Distributed by Belknap) Louisville, KY 60
O'Neil Mfg., West New York, NJ 56–58
Operetta (Victor Radio Corp.) Chicago, IL 59
Ortho Cone, Chicago, IL 60

P

Pacent Radio Corp., New York City, NY 61
Pal Radio Co. Inc., Jersey City, NJ 62
Pathe' Phono and Radio Corp., Brooklyn, NY 63–65
Pausin Engineering Co., Newark, NJ 69
Peerless (United Radio Corp.) Rochester, NY 66
Penn 67
Pennsylvania (Distributed by Hamburg Bros.) Pittsburgh, PA 67
Philco (Philadelphia Storage Battery Co.) Philadelphia, PA 68

Q

Quam Radio Co., Chicago, IL 70

R

RCA (Radio Corp. of America) New York City, NY 71, 72
R.F.I. (Radio Foundation Incorporated) New York City, NY 73, 74
Radialamp (Radiolamp Co.) New York City, NY 118
Rola Co., Oakland, CA 75

S

SAAL Co. H. G., Chicago, IL 76
Safar, Italy 77
Sandar Corp., Long Island City, NY 78
Sea Tone 118
Selector Co., Seattle, WA 79
Senesco, Boonton, NJ 79
Ship Horn Speaker (Fictitious name. Unknown manufacturer) 119
Ship Speaker (Fictitious name. Unknown manufacturer) 119
Silvertone WLS, (Distributed by Sears Roebuck Co.) Chicago, IL 80, 81
Slagle Co., Fort Wayne, IN 79
Sonochorde (Boudette Mfg. Co.) Chelsea & Revere, MA 82–84
Sparton (Sparks-Withington Co.) Jackson, MI 85
Splitdorf Radio Corp., Newark, NJ 86
Stevens & Co. Inc., New York City, NY 87
Sturrup-Plimpton-Newell Co., Worcester, MA 85
Stewart-Warner Speedometer Corp., Chicago, IL 88
Stromberg-Carlson Telephone Mfg. Co., Rochester, NY 89, 90
Symphonic Sales Corp., New York City, NY 91

T

Teletone Corp., New York City, NY 92
Temple (See Airchrome) 8
Thermatrol Mfg. Co., Springfield, MA 93
Thorola (Reichmann Co.) Chicago, IL 94
Timmons Radio Products Corp., Philadelphia, PA 95, 96
Tower Mfg. Corp., Boston, MA 97–100, 118
Trimm Radio Mfg. Co., Chicago, IL 101–103

U

Utah, Salt Lake City, UT 104–106

V

Vemarc (Associated Radio Corp.) Ann Arbor, MI 107
Victor Radio Corp., Chicago, IL 109
Vitalitone Radio Corp., New York City, NY 107, 108
Voice from the Sky (Florentine Art Productions Inc.) Chicago, IL 118

W

Walcone 110
Western Electric, New York City, NY 111–114
Westingale Electric Co., Chicago, IL 115
Westinghouse, Canada 115
Windsor Furniture Co., Chicago, IL 115, 116

Y

Yar-Lange (YAR), Milwaukee, WI 119